ENERGY
THE NEW LOOK

ENERGY
THE NEW LOOK

BY MARGARET O. HYDE

McGRAW-HILL BOOK COMPANY

New York • St. Louis • San Francisco • Auckland
Bogotá • Guatemala • Hamburg • Johannesburg
Lisbon • London • Madrid • Mexico
Montreal • New Delhi • Panama • Paris
San Juan • São Paulo • Singapore • Sydney
Tokyo • Toronto

TO JULIE ANN, BARBARA JEAN,
MARGARET MARY, AND
AMY KAY LIMOGE

Library of Congress Cataloging in Publication Data

Hyde, Margaret Oldroyd, 1917–
Energy, the new look.

Bibliography: p.
Includes index.
SUMMARY: Discusses certain sources of renewable energy which may be tapped in the future to help reduce dependency on fossil fuels.
1. Power resources—Juvenile literature.
[1. Power resources. 2. Renewable energy sources]
I. Title. TJ163.23.H92 621.042 80–21376
ISBN 0-07-031552-3

23456789BKPBKP876543

CONTENTS

1
LOOKING AT CHOICES

What is the energy scenario for tomorrow's world? Will you live in a home that is buried in the ground for temperature control and depend on solar batteries to supply the energy for your electric lights? Will your car be powered by the energy from cattails, or will you tap a tree in your garden for the diesel oil to fill the tank?

The blades of a wide variety of wind machines are already turning in unexpected places like the Bronx, New York. Waves are producing small amounts of electricity at sea, and solar-power towers are turning water into steam in semi-arid regions. Wood chips provide the heat for electricity in a city, and pig manure heats the home of a farmer. Fuel is being squeezed from rocks in increasingly larger amounts, and energy is being produced from the splitting and fusing of atoms. Certainly, there is a new look in the world of energy.

Few decisions in the present time are more important than those concerning the development of energy for the future. Such decisions will affect the shape of society. For example, the amount and results of current research will affect the quality of life in the future. Since energy-development projects commonly take from three to twelve years from the time the decision to proceed is made until they make their first sizable contribution to the supply, awareness of the needs of tomorrow must bring action today.

Choices must be made in the transition from the major sources of fuel in today's world: oil and gas. Hydropower, which is usually considered a clean and economical energy source, may supply as much as 3 percent of the energy in the United States by the year 2000. While many small hydroelectric plants may add something, most of the sites for large hydroelectric dams have already been developed. Forecasts for nuclear-power growth vary greatly, depending on whether the forecaster supports or opposes this energy path.

What are the choices among the new energy sources? At present some use more energy than they produce, but this may not always be the case. A few of the most exciting alternative sources of energy may soon move to a practical ratio of "energy in versus energy out." All exact some toll on human health and the environment, but some impose risks that are less severe than others. Some choices are far more practical economically than others.

Certainly, the world is not running out of energy, but the supply of inexpensive oil and gas has gone. Nonrenewable resources must give way to sources of energy that will never be depleted. Many of the ideas that sound impractical today may well be the basis for abundant energy tomorrow, but the road to renewable sources is hard and long. The sun does not always shine on the solar panels that are carefully placed to collect its light. The wind does not always blow as hard as necessary to turn the blades of the wind machines. The world is full of hydrogen as part of water, but the hydrogen is not easily separated from the water, nor is it easily stored. At the dawn of the nuclear age, energy from atoms was expected to usher in an era of abundant supply. Many scientists promised that electricity would be so plentiful and so cheap that meters would be eliminated. Now the risks and uncertainties of nuclear power are the subject of much controversy.

Are synthetic fuels from coal, shale oil, and tar sands the answer? Should the use of coal be greatly increased? Should the greatest emphasis be placed on developing solar energy? There is no one answer. Tomorrow's energy supply must come from many different sources, and every possible source must be explored and tested. At the same time, the effect of each on the planet's fragile environment must be carefully weighed.

Is the large-scale use of coal more dangerous than nuclear power? Some experts think it is, while others disagree. Can organic wastes be safely converted into fuel? How do the benefits of each potential source of energy compare with the risks?

Some sources of energy are more suitable for one kind of use than another. Consider how energy is used in the United States: 25% for electric generating plants, 25% for moving people and materials, 40% for industry, and 10% for a variety of other uses. About 90% of today's energy demands are met by fossil fuels—coal, petroleum, and natural gas. How can this amount be reduced?

You can join the search for answers to these questions. In the following pages you can look at some of the choices and consider the paths that you think should be explored further.

2
NEW FUELS FROM PLANTS

Picture a large rest area and refueling station on an express highway. One car goes to the pump for diesel fuel, and another fills up with gasoline. Some of the cars use gasohol to power their engines. There is one pump for pure alcohol fuel made from corn. Gasohol is part gasoline and part alcohol; the alcohol is made from plants such as corn that gather energy from the sun. Of course, all of the fuels, gasoline and diesel included, were originally living matter and in a true sense might be considered a form of solar energy. New kinds of fuels from plants are made mostly from fast-growing renewable sources.

"Biomass" is actually a general term for all animal and plant matter, but it is often used to mean plant matter that is used as fuel. In addition to wood, there are many kinds of biomass that are alternative sources of energy. Of course, in photosynthesis, green plants use solar energy to convert carbon dioxide and water to carbohydrates and oxygen. This process has puzzled scientists for many years, although a lot of their questions have now been answered. It has long been known that plants' ability to convert solar energy (sunshine) to stored energy (biomass) is very low. However, there is considerable variation among plant species in their efficiency of conversion. In the search for new fuels, energy-efficient plants are receiving new attention.

Most of the sunlight that falls on plants is reflected back or dissipated into the atmosphere. In some crops only 0.1 percent of the solar energy is captured, but in others the rate of efficiency is 3 percent or even better. Crops that use energy so efficiently might well be extremely helpful in supplementing other sources of fuel. An energy farm the size of the state of Texas could theoretically supply all the energy needs of the United States over a period of one year. But Texans, or people in any other state, are not about to give up their land for such a use.

Certainly, one of the problems in using plants for fuel is the amount of land they take from other uses. Will farmers find it more economical to produce food or fuel? Will there be energy farms on large areas of land that are not fertile enough for food crops? In some areas, forests of trees have long been serving as energy farms by supplying wood for fuel.

What other plants can help to supply fuel for today and tomorrow? Many people are saying, "If it grows, convert it to energy." One of the most appealing ways to use biomass to stretch the fuel supply is that of gasohol. Although supplies of gasohol may be limited and regional until 1985, major new facilities are producing this kind of fuel. Gasohol can be made by moistening plant matter, fermenting it with yeast, and then distilling it and dehydrating it to produce water-free alcohol.

Gasohol can be made from a variety of substances. Many of the first supplies came from corn, which was converted to ethyl alcohol and combined with gasoline in a ratio of ten parts ethyl alcohol to ninety parts gasoline. The type of alcohol known as ethyl can be made from various grains and from some other kinds of plants. Considerable research is being done around the world on the practical production of alcohol for gasohol. In Brazil, for example, sugar is being cultivated for alcohol production in a national energy program.

Research into the production of another kind of alcohol, known as methanol, indicates that it, too, might be substituted for gasoline or used in combination with it. Methanol is relatively easy to make from a variety of substances, such as coal, wood, garbage, and animal waste matter.

Methanol and ethanol alone or mixed with gasoline are popular fuels in many agricultural areas because the cost of transporting fuel to the pumps is low. In São Paulo, Brazil, alcohol is plentiful because of the abundance of sugar cane that is converted to fuel. Brazilian car owners can buy a kit to convert their gasoline engines for alcohol consumption. Some new cars have been redesigned for the use of pure alcohol as fuel.

Research from all over the world indicates that alcohol fuels, whether mixed with gasoline or pure, produce fewer pollutants than gasoline alone. Claims differ as to whether they are more or less efficient as far as gas mileage is concerned.

It is important to use energy-efficient crops to produce gasohol. There are many crops that convert solar energy more efficiently than corn, and many areas where other crops can be grown more abundantly. For example, in Hawaii, fuel prices are high, and sugar cane is abundant. Here and in other places pineapples, sorghum, and cassava are plants that may play an important role in the production of alcohol for fuel. Many experiments are under way to increase yield through the use of new strains, denser plants, and more efficient ways of converting crops to alcohol. Alcohol can already be produced from biomass cheaply enough to compete with gasoline as a motor fuel in some parts of the United States.

New processes that convert the *in*edible parts of plants to alcohol make the idea of plants for a transportation fuel even more appealing. Scientists have known for nearly two hundred years that cellulose, the basic building block of plants, can be converted to sugar and to alcohol. Before energy costs soared, there was little incentive for doing this on a large scale. Now new techniques for turning cellulose into alcohol are being explored and may make it possible to produce alcohol at a much lower price.

Since the strongest argument against the use of plants for energy is based on the scarcity of land that can support crops, scientists have looked to the sea for algae, in the form of kelp. For example, these large marine plants are being farmed on open ocean rafts off the coast of California. When dried, they can be converted to methane gas, which is a combination of carbon and hydrogen and is the major constituent of natural gas. Methane can be burned or converted to methanol, the kind of alcohol mentioned earlier.

Planners envision thousand-acre ocean farm projects, with processing plants, storage spaces, living quarters for workers, and helicopter landing pads. Since nutrients are in short supply in surface waters, the plan calls for pumping water from deeper layers to supply the nutrients that would stimulate the rapid growth of the kelp. Plastic lines would extend from the central unit far into the ocean to support the kelp. Since large amounts of solar energy fall on the ocean and are captured by the kelp, such a farm would be an exciting contribution to future energy programs. The kelp is fast-growing, does not take farmland away from the production of food, needs no irrigation, and even adds to the supply of fertilizers and animal feeds. But this kind of energy farm is still highly experimental.

The use of single-celled algae that grow in profusion in fresh water is also being explored. As in the case of kelp, harvesting and drying these algae is expensive, and there is an added problem here of supplying nutrients from the land. Growing the algae in sewage sludge is one possible solution to the nutrient problem. This attempt to use algae to add to the fuel supply is also highly experimental.

Water hyacinths have long been a nuisance plant. They have clogged drainage ditches and inland waterways, and they continue to be a problem in many areas because they grow so profusely. However, they are becoming attractive as a possible source of fuel. In some experiments where they have been grown in sewage, the nutrients on one acre produced more than a thousand pounds of dried water hyacinths. This amount could be used to generate about five thousand cubic feet of methane gas. It is estimated that over a billion cubic feet of methane gas per day could be harvested from the water hyacinths in the state of Louisiana alone. Even though harvesting these plants is expensive, they may someday be a practical alternative source of energy.

Cattails have been considered a nuisance weed

in recreational lakes and breeding sites for waterfowl, but energy experts are looking at them now as a possible source of fuel. This prolific water plant is one of the most efficient natural converters of solar energy ever investigated. Scientists Eville Gorham and Douglas Pratt at the University of Minnesota are exploring the energy potential of cattails. They believe cattails could be harvested in strips without disturbing the environment, fortunately for wildlife. The plants spread with underwater stems, called "rhizomes," that produce new shoots. When plants are harvested, these stems remain intact and plants can grow again.

Fossil fuels take thousands or millions of years to form, but these plants are renewable every year. Cattails may be a good crop to plant after peat lands are harvested for energy. Peat is the earliest stage of transition from compressed dead plants to the formation of coal, and it is found in places such as Ireland, Scandinavia, parts of the Soviet Union, Michigan, California, and the Florida Everglades. Peat takes about three thousand years to form and can be harvested only once, but cattails could be grown on a thin layer of the remaining peat and be a renewable source of energy. Cattails grow wild in many areas where the soil is very wet, in areas known as wetlands.

Although cattails burn fast and the rapid heat cannot be captured, they could be compressed into fuel pellets, or their starch could be converted to alcohol that could be used as fuel. Drs. Pratt and Gorham believe that cattails may be good supplements to the general energy supply in areas near where they grow. Minnesota alone is estimated to have ten million acres of wetlands that are suitable for their growth, and there may be 140,000 square miles of wetlands in the United States.

One of the most appealing sources of energy is the diesel fuel that grows in certain trees. As long ago as 1976, Nobel Laureate Melvin Calvin of the University of California at Berkeley became very interested in some trees in Brazil that produce significant amounts of an oily sap. In several varieties of trees belonging to the genus *Euphorbia,* hydrocarbons (oil) are produced instead of carbohydrates. For a long time the natives of the forest have used the oil from at least one variety of Euphorbia by drilling a small hole into the trunk of each tree and putting a bung, or plug, into it. About every six months they remove the bung and collect fifteen to twenty liters of the hydrocarbon. Since these people do not drive diesel cars, they use the oil as a healing ointment, a base for perfume, and in other ways.

Recently, on a visit to Brazil, Dr. Calvin reported that the Brazilians had tried the sap from such trees in the fuel tank of a car with great success. He suggests that many thousands of acres could be put to the production of these trees in the

United States, even on arid and semi-arid land that is not useful for food or fiber production.

Euphorbia lathryis, one kind of tree that pours out pure diesel fuel, is being grown at the South Coast Field Station of the University of California in Santa Ana. Here the plantation produced more than ten barrels of oil per acre in only seven months. Dr. Calvin reports that the caloric value of the oil grown in the trees compares favorably with that of crude oil. Certainly, this is a source of energy worth exploring further. Now that hydrocarbon sources are being depleted in the mines and wells under the soil, one might consider these trees a kind of oil well growing above the soil.

These are just some of the ways that experts are looking to plants growing on the land and in the water as possible supplements to the supply of usable energy.

The size of the role that biomass conversion will play in the future depends upon many factors—political, economic, scientific, and technological. Reductions in the cost of the conversion process through technological improvements are already helping to increase the potential for the use of new kinds of plants for fuels. While no one single plant is expected to be a major source, at least several have the potential to make good contributions to the energy future.

3
HARNESSING HEAT FROM BENEATH THE EARTH: GEOTHERMAL ENERGY

Today, a woman who lives about two hundred miles from Manila in the Philippines holds her pot of food over a hole in the earth that is hissing with the boiling of water far below the surface. Another woman steams a chicken in a hot rock crevice in her backyard. These women are cooking with heat from deep within the earth. In the area where they live, it has found its way to the surface naturally. Engineers work near the native women on a large generating plant that supplies some of the electrical power used in the city of Manila.

Heat from deep within the earth is known as geothermal heat. There is plenty of it, but it is seldom as easily available as it is for the women described here. However, there are many places where this energy is being harnessed.

As long ago as 1904, Italian engineers in Ladarello, Tuscany, put some of the steam from beneath the earth to work. They captured the steam from hot wells and sent it through pipelines that led to electrical generators. Less than fifty years after they began this operation, geothermal energy provided enough electricity to power most of Italy's railways. Geothermal energy is still used in producing some of Italy's electricity.

In Iceland, geothermal energy has supplied the heat for buildings since the 1930s. In the capital city of Reykjavik, an elaborate network of pipes and conduits carries hot water from one hundred geothermal wells to 90 percent of the city's homes. The temperature of the water, which is under pressure in the wells, ranges from just under 100 degrees Celsius (212 degrees Fahrenheit) to 150 degrees C (302 degrees F). It is about 80 degrees C (176 degrees F) when it reaches homes 16 kilometers (10 miles) away. In addition to heating buildings in Iceland, geothermal energy heats places such as swimming pools and greenhouses. Icelanders, descendants of the Vikings, frolic all winter in their open-air swimming pools, proudly patting snow on their bodies before and after plunging into the steaming waters for a swim. The greenhouses furnish fresh vegetables and flowers the year round.

Japan has made wide use of geothermal energy for hot water spas and greenhouses. This country also uses it for some commercial baking, experimental fish farming, sea-salt recovery from sea water, and in other ways.

In Paris, France, several blocks of apartment houses are heated by geothermal water that is extracted from depths of less than two thousand meters. There are plans to heat half a million dwellings in the Melun region of France with this method by the year 1985.

Geothermal energy is used over a wide geographical area in the U.S.S.R., furnishing hot water and heat for communities of thousands of people and also for some industrial purposes.

There are many other countries in which geothermal energy is being used to heat hot-water baths, houses, schools, hospitals, greenhouses, and other buildings. But heating space such as this is practical only when the areas are fairly near the hot wells that provide the energy.

There is more flexibility in the case of production of electricity from geothermal sources. The United States, Iceland, Italy, Japan, Mexico, New Zealand, and the U.S.S.R. are examples of countries that produce some electricity from heat deep within the earth.

Scientists estimate that some fifty countries around the world could be adding to their energy supplies by the use of geothermal stations. These countries are located in the areas that geologists call the "Ring of Fire." There, volcanoes form a belt which extends along the west coast of South America, the west coast of North America and Alaska, and through the Kamchatka Peninsula, Japan, the Philippines, Indonesia, and into Southern Asia. But even though the heat source is free and near the surface in many areas, putting geothermal energy to use may require money, imagination, and effort.

Harnessing the heat of the earth is certainly not easy, but it is worth the effort since the supply may be inexhaustible. Deep in the center of the earth, a few kinds of elements are believed to be decaying through the process known as radioactivity. As new elements are formed, energy is given off. That energy is changed into heat energy, which keeps the core of the earth hot and molten. Heat energy is radiated outward so that the crystalline rocks above the core are heated. Heat moves upward to a layer of porous rock which contains water that has percolated down from the ground or surface of the earth. At the level of the crystalline rocks, the water is heated and it is under pressure, a condition that keeps it liquid above 100 degrees C (212 F). Water and superheated steam fill the cracks and pores in the rock. At hot springs, geysers, and volcanoes, steam spurts out from the earth. The woman in the Philippines who was mentioned on page 14 was able to cook food over such steam. Near her, and in many places on the earth, hot rocks are exposed where they have been thrust to the surface of the earth.

At the surface, or relatively near it, there are formations such as dry steam fields and hot water reservoirs that can be used to harness the energy deep within the earth. Dry steam fields, or areas where hot water vapor is dominant, are the most attractive commercially, for in these the steam can be used as it is to power turbines that produce electricity. The Geysers, the Pacific Gas and Electric Company's power plant ninety miles north of San Francisco, is of this type. Here electricity is produced that helps to turn on the bright lights of the city of San Francisco and other Bay Area communities. It is hoped that The Geysers will be able to supply two million kilowatts of electricity by the year 1990, or enough to meet the needs of two million people.

Hot water reservoirs are much more common than those of dry steam. A geothermal test facility in California's Imperial Valley is now experimenting with energy from hot water reservoirs.

Geothermal energy is harnessed by drilling wells in areas where steam or hot water is believed to be near the surface. Hot water that is in the ground bursts into steam when the pressure on it is reduced at the surface of the earth. The steam is carried in pipes and used to turn turbines that generate electrical energy. In another kind of system, hot water is brought to the surface of the earth under pressure at its original temperature and in its liquid condition. This water is pumped through heat exchangers where the heat changes another liquid to gas, which, in turn, drives a turbine. In both cases, hot water is returned to the underground reservoir.

According to experts at the Union Oil Company, a pioneer in developing geothermal energy, there is the geological opportunity to develop twenty million kilowatts of geothermal generating capacity before the end of the century. It would take 700,000 barrels of oil per day to generate this much electricity.

One of the advantages of using geothermal energy is that there are very few environmental problems associated with it. One hazard, however, is that the land might sink at places where large amounts of water have been drawn from beneath it. But this kind of situation may be prevented by withdrawing hot water at a rate that is safe and by re-injecting cold water into the wells.

Much progress has been made in a very short time in establishing geothermal energy as a useful contributor to the world's supply. The potential for the future is still being explored. In the United States, the best prospects are in the states west of the Great Plains from Canada to the Mexican border. Some areas along the Gulf Coast also might benefit from this source of energy. Geothermal projects are scheduled, are under way, or are in action in New Mexico, Utah, Idaho, Nevada, Oregon, and California.

Research continues on a number of fronts. For example, taking heat from hot rocks that lie near the surface may supply a significant portion of geothermal energy in the future if extracting the heat from them can be made practical. Electricity-producing turbines surged to life in the spring of 1980 powered for the first time in history by the heat of dry, hot granite. In an experiment conducted by the University of California's Los Alamos Scientific Laboratory, some of the power for a local research site was fueled by water heated deep within the earth in the Jemez Mountains of north-central New Mexico.

In the dry, hot rock approach to geothermal energy, two holes are drilled into the relatively impermeable rock deep in the earth. Water is forced down one hole with enough pressure to crack rocks between the twin holes, thus creating a heat reservoir. Then, fresh water is circulated through the injection hole, heated in the reservoir of hot rocks, and brought to the surface at a temperature well above the boiling point at ordinary pressures. The water is held under enough pressure to keep it liquid and is used to spin turbines that produce electricity. Energy gained this way can also be used for heating buildings and for industrial processes.

Although there are more technical problems with using dry rock, there is evidence to indicate that the total geothermal energy in dry rock is much greater than that generated by hot water and steam systems.

Another idea is to tap directly into molten rock, or magma, near a volcano. A heat collector would be used to bring the heat to a power plant at some distance from the volcano area. The temperature of the magma fairly near the surface of a volcano might be 1000 degrees C (1800 degrees F), or even higher. A cubic mile of such material is estimated to contain enough heat to run several electric power plants for a hundred years. However, many technical problems must be solved before such a scheme could be attempted.

Still another approach to taking energy from deep in the earth would make use of the steam that is known to exist at depths of about 1500 to 7600 meters (5000 to 25,000 feet) along the Gulf coast of Texas and Louisiana. There is evidence that steam (also known as "brine" because of the chemicals dissolved in it) can also be found in other parts of the world at these depths. Steam at these depths is often referred to as "geopressured." Researchers at the Southwest Research Institute, San Antonio, Texas, hope to determine how existing technology can extract and use the energy from geopressured steam to produce commercial electrical power. They hope to demonstrate whether or not this is a practical energy source by the mid-1980s. With current technology, drilling below 6000 meters is not routine.

On a world scale, the amount of geothermal energy now produced is relatively small. No one knows how much it will contribute in the future. Ideas that were once considered unrealistic are now being taken more seriously because of the increased need for new sources of energy. For example, some experts believe that more research may make it possible to extract energy from hot dry rocks at costs that are competitive with nuclear-fueled or conventional plants.

The amount of geothermal heat has been calculated to a depth of six miles for the whole United States. If it could be brought to the surface, there would be enough to heat all the homes in the country for thousands of centuries. But no one knows how many places throughout the United States are hot enough to be of practical value.

Some experts predict that it may supply about 2 percent of the total electrical generating capacity in the United States by the year 1985. Some predictions are more optimistic, and others are less so. One United Nations survey suggests that if all the geothermal energy available in Ethiopia were developed, it could satisfy the present electrical needs of all of Africa.

The untapped energy beneath the surface of the earth in the United States may be five thousand times greater than the nation's current annual energy consumption. However, many technical and economic hurdles must be overcome. For example, the problem of corrosion from the chemicals that are released with the hot water or steam presents a difficult challenge.

Since the breakdown of radioactive elements will continue to heat the earth's core no matter how much heat is used for geothermal energy, the source can be considered virtually inexhaustible. Certainly, harnessing the natural heat of the earth has important potential in America's energy picture.

4
HARNESSING THE WIND

The use of wind energy is old indeed. Long before the picturesque windmills of Holland were built, sailors used the force of wind to drive their ships. The ancient Egyptians depicted a sailing craft in drawings that date back to 3000 B.C., and scholars say that Buddhist monks near Tibet have relied on the wind to turn their prayer wheels for 2500 years. In the ancient land of Persia (which later became Iran), scholars have found the remains of wind machines that date back to early times. Here the wind was used to turn an eight-bladed rotor with wind sails woven from reeds. The bottom of the windmill's shaft turned a millstone to grind grain into flour. By the twelfth century, windmills that ground grain were common in Europe. Later, windmills that were used for pumping water and for irrigation became popular.

The famous Dutch windmills were erected to carry out large-scale projects of land reclamation—draining water from land so it could then be used for agriculture. In some places the force of wind on the blades lifted the water as much as twenty feet and then channeled it elsewhere. Drainage projects were begun in the early seventeenth century, and by the nineteenth century there were nine thousand Dutch windmills. England had almost as many windmills for drainage and flood control. When electric and diesel power took over, the windmills almost everywhere fell idle. They were abandoned to decay, or were dismantled. However, picturesque canvas sails still turn in the wind in the Mediterranean area and many other parts of the world.

In the United States, about six million multibladed windmills were built and used to pump water on American farms before the rural electrification program that began in 1930 and extended through the next twenty years. In a speech, President Carter reminded Americans that there were windmills on the Great Plains long before there were high-tension wires.

A windmill of today may bear little resemblance to the picturesque windmill of yesterday, although each is not without its own appeal. The world's first large wind turbine operated between 1941 and 1945 at Grandpa's Knob near Rutland, Vermont. This windmill was designed by Palmer Cosslett Putnam, who envisioned as many as one hundred

of these large windmills to be placed strategically in windy spots on the New England Power network so that they could help to reduce the consumption of fossil fuels during peak-load periods. The wind machine's propeller was connected to an electric generator that produced a sizable amount of electricity. Unfortunately, complications arose that prevented the extended program from being put into action. For example, in 1943 a bearing failed, and the machine had to be shut down for months. When one of the eight-ton blades was found to have a flaw, it could not be replaced because of wartime priorities. So the wind machine ran with the defective blade until one night in March 1945 when the blade fell off and tumbled seven hundred feet down the mountainside. Still, the Vermont experiment gave ample proof of the practicality of wind-produced electric power.

Serious interest in wind machines has been renewed, but at a pace that enthusiasts consider very slow. New models are being developed with the United States Department of Energy funding, and new-generation wind turbines are being marketed for private use. Both large-scale and small-scale systems are being developed, with many new designs competing to determine the most efficient. By the fall of 1979, nineteen utilities companies had applied to the United States Department of Energy for designation to build windmills as part of the Federal research program. One, for example, was the Long Island Lighting Company in New York, which proposed building two windmills that would rise 350 feet and support 300-foot blades. These could generate enough electricity to amount to about 0.25 percent of the utility company's output. The mean constant year-round wind in that area has been calculated to be 13.4 miles per hour, just about the minimum amount needed to make the windmills practical.

Many people are suggesting the construction of small windmills in suburbs to supplement the supply of fossil fuels. Unfortunately, getting electricity from windmills is more complicated than putting some blades on a tower, for not all places have enough wind. If you wanted to build a windmill near your home, the first step would be to measure the wind for a year with an anemometer, which records the speed of the wind at that location. A difference of one or two miles per hour in the annual average wind speed can make a difference between a wind machine that pays for itself in a few years and one that is highly impractical. If the spot proves windy enough, you could proceed with investigations of designs, methods of storing energy when the wind is not blowing, and ways to connect the energy from the wind machine to the commercial supply.

In the United States there may be about 1000

wind machines tied to utility company lines; the customers send power back to the utility companies when they have excess. Two-way electric meters keep track of what is going in each direction, but there are many regulations for the homeowner or small group that wants to use wind energy in conjunction with an existing utility service.

The blades of a wide variety of wind turbines are turning in some of the most unexpected places. For example, in the fall of 1979, New York City's first commercial-scale windmill in more than two centuries provided the power for a sound system during the ceremonies for the windmill's dedication. This wind turbine was named Aeolus after the Greek god of the wind. It was built by community action and erected on Hunts Point peninsula in the Bronx, where the windmill generates forty kilowatts of power. Batteries store as much as a hundred kilowatt hours, and excess energy is fed into the Consolidated Edison grid for a fee. Income from excess energy is budgeted for the nonprofit Bronx Frontier Development Corporation for a program to supply fertilizer for local parks and gardens. These projects resulted from the work of citizens who wanted to make the dumps and abandoned lots in their communities green and attractive.

Small wind systems vary from those that produce only enough electricity to power small appliances to those that produce close to 100 kilowatts of electricity. The average American home uses approximately 700 kilowatt hours a month, or an average of one kilowatt (1000 watts) per hour. At Rocky Flats, near Golden, Colorado, small wind-machine systems with blades from ten to thirty feet in diameter are being tested in the technical and economic aspects of their use.

The first major Federal project was a large experimental wind turbine generator built near Sandusky, Ohio, under the direction of the National Aeronautics and Space Administration. Known as Mod-0, this wind machine was built with two slender rotor blades like airplane wings, which span a distance of 125 feet and are set on a tower 100 feet high to catch the winds. When the wind speed reaches eight miles per hour, power is generated, but full power requires winds of eighteen miles per hour.

All windmills are situated high above the earth where the wind blows more strongly than near the ground, but they vary greatly in appearance. One variety, known as the Darrieus after its inventor, uses eggbeater-shaped blades and a vertical axis. This kind of windmill is not self-starting, but it does work with wind from any direction. A large wind machine of this type has two thin blades each only two feet wide on a wind machine that is 120 feet tall and 80 feet in diameter.

A number of giant wind machines of the propeller type have been built at windy sites in various parts of the country.

Generally, people welcome "free energy from the wind." For example, at Boone, North Carolina, people celebrated the dedication of their new giant wind machine with the waving of pinwheels and the release of hundreds of colorful balloons into the air. This $5.2 million wind turbine is one in a series of experiments sponsored by the Department of Energy and the National Aeronautics and Space Administration. A year after the Boone windmill was erected, a wind generator with twice the electrical generating capacity was taking shape near Medicine Bow, Wyoming. This was the first of fifty windmills planned for that area.

While the wind is free, wind machines are not without their problems. For example, the people of Block Island, Rhode Island, have been plagued by the unbearable noise produced by their windmill, blades that blew off, and a bent shaft. The wind is free, but the cost of the machines is usually one of their disadvantages. The higher the tower, the more wind, but the height of the tower is a major factor in building costs. To make the wind system economical, blades must last twenty-five to thirty years, but some of the experimental blades have cracked and blown apart in gale winds when they were quite new. When blades can be mass-produced and structural difficulties overcome, wind energy will become less expensive.

This energy source seems to generate no pollution other than noise but the unsightly appearances of the mills have been the cause of some complaints. Heavy metal blades on windmills have interfered with local television reception so much that cable systems have had to be installed to overcome the complaints of viewers. If large numbers of small windmills are combined to produce power, they must compete for land. The storage of wind power is another problem because the supply of moving air is uneven.

As research continues, some of the disadvantages may be lessened. Wind energy has many advantages. This clean and plentiful source does not stop at night or on cloudy days when sunshine is blocked from parts of the earth. The planet's atmosphere acts like a giant heat engine, with hot, tropical air rising and cooler polar air moving in to replace it. This flow is modified by the rotation of the earth on its axis, which produces the familiar trade winds, and by local variations that lead to changes in wind speed and direction. But the sun will continue to stoke the earth's huge thermal engine no matter how many wind machines take their energy from the air that blows across them.

Will the wind be harnessed once more for ocean travel? Can sailing ships carry cargo less expensively than ships powered by high-priced and scarce oil?

Some experts believe that wind energy might

play a major role in powering the cargo ships that now use about 100,000 barrels of oil per day. Sail power has the potential ultimately to fulfill half or more of all ocean transport requirements.

While some people scoff at the idea of sail power for the world's cargo ships, others are busy gathering data that compare the cost of a sail-power system with the cost of a fossil-fuel system. Some rough calculations were made as long ago as 1976, when the cost of fossil fuel was much lower. They indicated that the average cost per year of a sail-power system, including the original investment, would, over a twenty-five-year period, be only 15 percent of the cost of the energy required for a ship powered by fossil fuel. This calculation was challenged, but accurate figures are not yet available. It is believed that square rigs designed with new materials and today's scientific knowledge would show a great deal of improvement over the performance of the old sailing ships. With modern marine engineering, commercial sailing ships could probably be greatly improved in many respects. And the use of sophisticated weather routing could greatly increase the voyage speed of sailing ships.

The new class of sailing ships would probably be manned by crews much smaller than the old square-riggers. Modern sails are made of materials that can be repaired relatively easily on board ship, saving much of the time that was formerly devoted to the making of new sails at sea and maintaining and repairing old sails on board.

Further research in using the energy from the wind to power cargo ships is under way. Researchers are testing the effect of modifying ship characteristics, and they are studying routes, safety features, weather variations, and operating costs in suitable computerized models.

The supporters of increased research suggest that a new class of jobs with unique appeal for sailing enthusiasts might be created. They believe that, with enough support for research, one might expect some full-scale trans-ocean ships to be sailing in commercial trade five years after the beginning of a suitable test program. Before long, wind energy could replace a significant portion of the oil used in sea transport.

5
SOLAR ENERGY

The sun shines brightly on an energy-hungry world. Actually, 99.8 percent of the earth's energy comes directly or indirectly from the sun. In a true sense, fossil fuels (oil, gas, and coal) retain energy from the sunshine of long ago. The wind blows because of the sun's uneven heating of the earth, so wind power is really sun power. Energy from plants comes originally from the sun. And so it is with all kinds of energy other than nuclear fission and fusion. There energy comes from the breaking apart or fusing of the nuclei of atoms.

Generally speaking, solar energy refers to the use of the sun's rays in direct systems. For example, flat-plate solar collectors are used for heating and cooling, focusing collectors supply heat to generate electricity, and solar cells, such as the silicon cells, convert sunlight directly into electricity.

Solar energy does not pollute the air or emit the kind of radiation that comes from nuclear power plants. It is without limit in quantity and duration.

The inexhaustible amount of solar energy that reaches the earth's surface in two weeks is equivalent to the energy in all known fossil fuel reserves. Why not convert everything to solar energy and forget about the fuel crisis?

The energy sources suggested in a plan for a "Solar Sweden by 2015" include windmills, hydroelectric plants, biomass (organic matter) from plantations on land and in water, and solar cells. Dependence on renewable energy forms would climb from 24 percent today to 50 percent in the year 2000 and 100 percent in 2015. The reality of this depends largely on cost and available labor.

Almost everyone agrees that solar energy can play a large part in the future, although not many people in the United States expect to live in a solar-powered utopia. Estimates of when and how much solar energy we'll use vary greatly, but many homeowners are already using solar energy in a small way to supplement their other methods of heating and cooling. Each small contribution helps both the individual and the community at large.

Perhaps your neighbors heat their water with solar panels on the roof of their house. Another neighbor might heat the entire house by tapping the energy from the sun with solar panels on the roof and a storage unit in the basement. The farm down the road may depend on solar energy for irrigation of crops. The fountain in your nearby park may be powered by solar cells on the lawn in an experiment sponsored by your local electric company.

Hot-water heaters are practical in many areas where there is not enough sun for space heating. The heating of water for everyday use is the most popular use of energy from the sun.

Two distinct methods are important in space heating and cooling, and they are known as active and passive. Active systems generally use mechanical power to distribute and store the energy. Passive systems use the natural flow of energy (conduction, convection, and/or radiation) and the buildings themselves to trap, store, and conduct the energy within the system. Some designs combine both active and passive systems.

Suppose your family is considering building a house that uses solar energy. Before going ahead you might visit an area where five houses illustrate different ways of making use of passive solar energy. The first house makes use of a system known as direct gain. This house has a large glass area on the south side and is well insulated in other areas. The sun shines through the glass and most of the light becomes heat that warms up the living space directly. Energy is stored in the concrete walls, slate floors, stone fireplaces, and some large containers of water on the porch and in the garage. If you have walked barefoot in the city in the summer, you know that concrete, stone, slate, and brick can become very hot from the heat of the sun.

Water, too, is a good medium for storing the heat of the sun. Heat stored in these materials helps warm the house at night and on cloudy days. Draperies and shutters close off the glass area at night to help retain the heat.

The second, House II, has a thermal-storage wall. This house was designed so that there is a large brick wall directly behind a large glassed-in porch. The heat is absorbed and conducted through this brick wall. Vents at the top and bottom of the thermal-storage wall permit some of the heat to flow directly into the building. Thermal walls can be made of other heavy materials such as concrete, stone, adobe, earth, or containers of water.

House III depends on a system that is sometimes called indirect gain, and has a thermal-storage roof. Heat-storage materials are placed above the ceiling of the building in a horizontal position. Most of such roofs store solar heat in plastic pillows of water that are placed above the ceiling of the one-story building. Glass panels on the roof, which slants, allow the sunshine to enter and heat the thermal-storage roof. The building below is warmed by heat that moves downward from the ceiling. When night comes, and when days are cloudy, insulation is moved into place over the ceiling and/or under the glass in the roof to prevent heat loss.

House IV has a sunspace that is used for a

greenhouse and family room. This greenhouse extends across the south side of the house. The glass permits the sun to heat the concrete floor and the containers of water in the greenhouse which, in turn, heat the area when the sun is not shining. The curtains are closed to shut out the heat loss at such times.

House V makes use of a convective loop that depends on the fact that warm air is lighter than cold air. Picture a house with a solar unit attached. The unit consists of a slanting glass-wall collector, which is heated by the sun's rays, and an air space beneath it. Since the whole unit is lower than the house, hot air rises into a duct and flows into a bed of rock beneath the porch of the house. From here the hot air travels through another duct into the house. Cool air drops through a duct at the back of the house and is carried to the rock bed and the collector area, where it is reheated. The air circulates naturally.

Active solar heating systems are more complicated and may be more expensive than the passive type. The active systems use fans or pumps and collectors that usually consist of a black metal surface enclosed in an insulated box with a glass or plastic cover. Collected heat is transferred to air, water, or another liquid, and piped to a storage tank of liquid or rocks. There are many variations in materials and construction of collectors.

Relatively inexpensive collectors that are popular for heating the water in swimming pools consist of mats of synthetic rubber tubing that are placed on a roof or in a sunny area. The water in the tubes is heated by the sun and then circulated in the pool.

In designing solar heating systems, the hot season of the year must be considered. The simplest systems have insulators that keep out the heat when it is not wanted, and these are usually the same ones used to keep the heat in on winter nights and cloudy days. But in hot regions, cooling is an important use of solar energy. Most of these cooling systems are more complicated than those used for space heating; but they do work, and they make solar installations valuable year-round sources of energy. Although the initial cost of some systems is high, many users are considering the fact that they are paying for their heating and cooling bills all at once.

Solar energy may someday provide the electricity for the houses that get their heat from the sun as well as for many that do not. Unfortunately, solar thermal power systems, those that harness the energy from the sun to heat the steam for electric generators, are still very experimental.

High in the Pyrenees Mountains of France, scientists have been experimenting for many years with ways of concentrating the rays of the sun to produce extremely high temperatures. As part of these experiments, France began the operation of its first solar generating plant there in January of 1977. One wall of a building is concave in shape and is covered with reflecting mirrors that concentrate the rays of the sun on a boiler in front of them. The water in the boiler turns to steam, which runs turbines to generate a small amount of electricity.

Solar-power towers in sunny locations in the United States follow the same principle as the solar generating system in France, but they use large numbers of sun-tracking mirrors called heliostats that are placed on one hundred or more acres of land. The estimate for the land needed for the heliostats of a commercial-size plant is one square mile; the plant may use as many as 10,000 heliostats to collect sunshine. These mirrors focus sunlight at the top of a tall tower. One experimental tower is about twenty stories high and another about fifty stories high. For a commercial plant of this size, the tower may be a thousand feet high.

Storage systems of solar-power towers consist of materials that hold the heat. For example, the pilot plant at Barstow, California, uses a huge tank filled with oil and rock to store enough heat for hours of electricity generation. This plant design will be followed in many places if it is successful.

At present, solar-power towers are designed primarily to supplement the power needed during peak periods or intermediate periods, such as mid-day, when the sunshine is brightest and an intermediate amount of power is needed by consumers. Every additional use of energy from the sun saves fossil fuels.

The potential for solar energy from solar-power towers goes far beyond the production of electricity. The basic "product" is steam, and steam is considered an ideal energy-transfer medium. Steam can be used for many industrial processes, such as driving irrigation pumps, changing salt water to fresh, and refrigeration, as well as space heating and cooling in industrial, commercial, and residential areas. Although experimental projects are operating now, many experts think that solar-power towers may not be practical before the year 2000.

Even though the sunshine is pollution-free, environmentalists are concerned about the impact of large collectors in desert areas where scarce water may be needed in the power towers. There is the possibility of the heat at the top of the tower causing local climate changes. This heat could create strong local winds because the lighter hot air would rise and the cool air would rush in to replace it. The space for heliostats is also a concern. Solar-power towers have no major technical problems, but they are very expensive to build and maintain at the present time.

Of all the choices in the solar supermarket, the

one that seems most appealing is the conversion of solar energy directly into electricity by means of solar cells. In goes the sun, out comes electricity. The principles are well known, and the cells, called photovoltaic, have been used successfully since their introduction by Bell Laboratories in 1950. Silicon, the principal raw material used in their manufacture, is the most abundant solid element on earth, but production costs are still very high. When much energy is required for production, there may be no net energy gain.

Silicon crystals in the solar cells are semiconductors, materials that conduct electricity better than insulators, such as rubber, but not as well as conductors, such as copper or other metals. Single-crystal cells are usually three inches across, and they are more efficient than a combination of smaller crystals in converting the sunlight that hits them into electricity. But single crystals are expensive and time-consuming to make. Thin films of material are less efficient than single crystals, but they are less expensive to produce. Thin-film cells can be deposited on inexpensive materials and can even be applied to rooftops. Or they can be extruded, or thrust out of machines, in the form of ribbons 8/1000 of an inch thick.

No matter how the material in a solar cell is formed, the sensitive material absorbs sunlight (photons), and this absorption frees electrons (parts of atoms) and allows them to move. Thus, a flow of electrons, or electricity, is created.

A variety of semiconductor materials is being explored for use in solar cells, but the typical solar cell consists of two ultrathin and dissimilar layers of silicon with an outside wire attached. The layers differ because a very minute amount of silicon in one of the layers has been carefully replaced by phosphorus, and a very minute amount of silicon in the other layer has been carefully replaced by boron. When the sunlight falls on the cell, electrons move along the wire from the layer with the phosphorus "impurity" to the layer with the boron "impurity." The current is collected by a grid of contacts on the top and bottom of each cell.

Groups of solar cells are often placed into a sealed or plastic unit called a "module." Each of these modules might deliver twelve watts of electricity, and groups of modules are positioned in rigid frames to provide the desired amount. Such groups are called an "array." Arrays are arranged at a specific site to form a field, the number of arrays depending on the amount of electricity needed for a project.

Forty solar cells connected together provide enough electricity to charge an automobile battery. Large numbers of them produce enough electricity to power satellites, buoys at sea, radios in remote areas, and unmanned oil and gas platforms. Solar cells operate lights and waste-disposal systems in some park facilities where other sources of electricity are not available. They are providing electricity to a naval station in California, a Papago Indian reservation in Arizona, and to experimental projects in educational institutions elsewhere.

An 800-module solar array containing 33,600 photovoltaic cells began powering radio station WNBO in Bryan, Ohio, late in the summer of 1979. The commercial power connection that originally provided the station's power is used as a backup, but it is needed only occasionally. Batteries store power for cloudy days. At times there has been enough power for studios, newsrooms, and production rooms. A daytime radio station is an excellent application for solar energy, since most of the energy is needed when the sun is shining brightly. The use for this kind of project, and for many others, should increase when the price of solar-generated electricity decreases.

Research on many fronts is helping to lower the cost of solar cells, and great strides have already been made. As the cost of these cells declines, their use increases. Power for whole villages may follow the use of solar cells to supply electricity in areas where none is available today. Use may continue to grow until the contribution to the total energy needs of the United States and other countries is quite large.

The need for new sources of energy is so great that the Congress of the United States began serious consideration of plans for solar-power satellites in the fall of 1979. While many people call the satellite a trillion dollar pie in the sky and a blue-sky idea, others take the idea seriously. The Solar Power Satellite Program (SPS) would involve building a new rocket five times larger than the *Saturn V* that was used in Project Apollo. A powerful space tug would be needed as well, and two shuttle-type spacecraft to ferry engineers and technicians into low earth orbit to assemble the satellites. According to another plan, shuttles would carry engineers and technicians to an orbit 36,000 kilometers (22,000 miles) above the earth, where the SPS system would be assembled.

A satellite in this program would be immense, possibly the size of the whole island of Manhattan, with photovoltaic cells covering a fifty-five-mile surface. The cells would convert energy from the sun to electricity, and then convert electricity to microwave energy that would be beamed to earth. Microwave energy would be captured by special earth-based receiving antennas and then reconverted to electricity that could be sent to meet the needs of cities.

Many of the questions about the solar-power satellite system are concerned with how it would be built. According to one estimate, workers would have to be launched into space by about five hundred blast-offs every year for thirty years. How productive would such workers be when they are floating in space suits or space vehicles? Could the solar space array really be assembled in low earth orbit and sent into its orbit by an ion thruster? How much would this program cost? Could such satellites supply as much as one-third of the world's electrical requirements by the year 2025 if enough tax dollars were voted for it? Some supporters of the program think it could, but many questions remain unanswered.

Environmentalists question the effect of microwave radiation covering several miles of the earth's surface. What will the chronic health effects be from low-dose, long-term, nonionizing microwave radiation? What will this mean to air travelers?

What will be the effects on communications systems? Will there be military dangers?

Supporters of the solar-satellite program point out that the unknowns may be fewer and less potentially hazardous than those associated with nuclear waste disposal or the effects of increased amounts of carbon dioxide from the burning of fossil fuels. Certainly there would be many advantages to such a system. Energy from solar satellites would be supplied round the clock, since the sun never sets in space. Satellites would capture solar energy before large amounts were lost on the trip through the earth's atmosphere. And of course the fuel supply is good for at least six billion years, and it is free.

Making the "free" solar energy less expensive is a goal that applies to almost all the systems contemplated. The amount that solar energy will contribute to the total picture varies in estimates of time and degree.

In some areas such as southern California, solar electric power plants may be available as proven commercial technology by the early 1990s. The Solar Energy Research Institute at Golden, Colorado, does research and development and functions in other ways to support the National Solar Energy Program. It aims to help establish a solar-energy industrial base that will foster the widespread use of solar technology. Many researchers and supporters of solar energy are working toward the day when it will be used to generate large blocks of power to meet the needs of cities and suburbs, factories and farms.

6
WOOD AS FUEL

Gagan, whose home is in India, spends most of his day gathering wood for cooking fuel. He picks up every twig that he can find and carefully bundles all the little pieces together. The chore was less difficult a year ago, for he could find enough wood for his family's needs in an hour. Now he may not locate enough in a whole day.

Gathering firewood is more difficult today for the 90 percent of the people of the undeveloped countries of the world who depend on it as their chief source of fuel. This means that one-third of the people of the world, roughly one and one half billion people, depend on firewood. Since the population of human beings is increasing much faster than the numbers of trees in the forests of the world, wood has become scarce. Many of these people live in areas far from mass communications, and their firewood crisis is seldom noticed. Nevertheless, it is quite real, and its effects are far-reaching.

As forests of the world are destroyed for fuel, the land on which they grew erodes and becomes unproductive. The food supply decreases and the deserts spread.

In many parts of the world, wood is a renewable resource, and this could be true in many more areas. Improved forest management, the planting of trees on unused land throughout rural areas of poor countries, protection against disease and insects, the planting of superior seed, the creation of fast-growing species on energy farms, and better use of wood waste all play an important part in the amount of energy that can be supplied from wood both now and in the future. The United States Forest Service estimates that the average number of trees produced on all commercial forest land in the United States is only about 61 percent of what could be produced. But the amount of forest land keeps shrinking as land is turned into building lots, highways, and parking areas. Large amounts of forest land have been taken out of production through the creation of wilderness areas. Good forest management is vital if wood is to continue to be a renewable resource.

In the United States, wood was the nation's first important source of fuel, and as recently as 1900 wood supplied 25 percent of the country's energy needs. But in 1976, only 1.5 percent of the country's total energy came from wood. It may supply as much as 10 percent in the future, however.

Many of the people who now look to wood as a source of energy are thinking about it as if it were a newly discovered source. The forest resource is discussed almost as if it were a recent invention. Certainly it is now viewed by many as nature's own solar collector that endlessly converts sunlight to energy and stores it in a form convenient for heating and other uses.

A caveman who used the heat from burning twigs to warm his hands gave no thought to the connection between energy from the sun and the energy released in fire. Today the wood in trees is recognized as an orderly chemical structure built from substances in the soil, air, and water, with energy from the sun. Some of the chemical substances are released when wood is burned, and a part of the solar energy is extracted in usable form as heat. As the oil supply has diminished and increased in cost, many wood-burning stoves have been installed as part of the heating systems in houses that are also equipped with oil burners or other sources of heat.

Many forest industries have returned to the use of scrap wood for fuel, converting back from oil and gas, which were once inexpensive and plentiful. With present harvesting methods, about eighty-three million tons of wood are left to waste on the ground each year in the forests. Mills produce about twenty-four million tons of wastes. It has been estimated that the United States forest industry could become self-sufficient in energy by the year 1990 by using wastes alone.

Even with the constraints needed to make it a renewable resource, wood appears to have a rather large potential role to play in the energy future of the United States.

Unlike many other sources of energy, wood's role in the fuel supply depends partly on other uses that can be made of it. For example, a tree standing in the forest may be most valuable for its lumber, or it may be most valuable as a part of a wilderness area. Much wood is used in the manufacture of paper and other products. Managing the wood resource so that it will not be depleted is most important for the quality of life in the future. The United States Forest Service estimates that the nation's forests currently could yield as much as five hundred million tons of wood a year for fuel without decreasing the amount used for paper and lumber.

Many trees in the forest are known as "junk wood." Removing these for fuel and thinning young stands of trees is becoming more common now that there is a greater demand. Thinning reduces the number of trees per acre so that the remaining trees can grow faster, with less competition for nutrients, water, and light. The trees removed during thinning are usually the less desirable ones. In the past, thinning was not so common because the thinned trees were of no commercial use. Demand for fuel wood now provides a market for these trees.

Today trees are being harvested to provide energy for factories and to supply the fuel to produce electricity for large numbers of homes. In 1977 an aged coal-fired generator in Burlington, Vermont, was rehabilitated to burn wood chips. The Burlington Electric Department retrofitted the generator with a do-it-yourself approach, using old parts, such as recycled steam pipes from other boilers, and about $25,000 in cash. Only a few weeks after the first wood chips slipped down the chutes to the boiler, popularly called Number One, it was discovered that this makeshift burner was generating commercial quantities of power, producing steam and driving the turbines. In the long term, the cost was lower than that of producing electricity with nuclear fuels in reactors, or by burning coal in conventional boilers. A second boiler was converted to wood-burning in 1979.

The process of providing electricity from wood fuel in Burlington begins with the harvesting of trees. They are fed into a machine that chops them into matchbook-size chips and then sprays them into trucks, each holding twenty-three tons. The chips are then transported from nearby forests to the city's generating station and are unloaded with a hydraulic truck dumper into the stockpile. From there they are moved by conveyor and deposited in a bunker high above the boiler. Machinery pulls the wood chips into troughs at a speed adjusted to the amount of fuel needed. From these troughs they are introduced into the wood-fired boiler. Four chutes carry the wood forty-five feet down the front of Number One. When the chips reach the boiler, they are blown into the hot fire by air that is forced through steel tubes. The steam that is produced is used to generate electricity just as in the case of electricity that is produced from any other fuel.

The 1977 wood-chip power project of the Burlington, Vermont, Electric Department was the first in the United States, and it was an exciting beginning. Here and in other areas, plans are moving forward and larger plants that produce electric power from wood are being put into action.

By the beginning of 1979, more than 150 industrial firms in New England had changed from fossil fuels to wood power. One large industrial plant that is planning to switch from coal to wood may burn 180,000 tons of wood per year. While this may mean a savings of 30 to 50 percent on the cost of fuel, many people raise the question about what heavy industrial use might mean to the supply of wood.

Certainly wood is not the ideal fuel for power plants and factories in all parts of the country, but it may be a good one in areas where forests are vast. While environmentalists are concerned about the possible depletion of forests, industrialists point out that they will draw on wood waste as well as on growing trees. In Vermont, where there are more trees than there were a hundred years ago, some forestry officials believe that the use of wood by the Burlington Electric Department could improve the quality of the forests through removing the defective and less desirable trees.

There is increasing concern about pollution from burning wood, however, as the amount used for fuel grows. The clouds of smoke that swirl from the chimneys of wood-burning stoves contain particles of soot and ash that irritate the eyes and dirty the environment. In recent studies for the Environmental Protection Agency, burning wood was found to emit far more particles suspected of causing cancer than oil or natural gas and about as many particles as the burning of coal. Also, wood gives off more carbon monoxide gas than oil or gas do.

On the average, about 80 percent of wood is consumed in the burning process, and the rest is left as ash or escapes up the chimney. In the case of oil and gas, 100 percent of the material is burned. In northern New England, where about 60 percent of the households heat completely or partly with wood, pollution caused by burning wood is a growing problem. In the communities that are located in valleys, such as Waterbury, Vermont, warm air can become trapped under a layer of cold air and the pollution problem can become serious. In this village, stacks of firewood lean against neat frame houses. In addition to the smoke from their chimneys, there is the smoke from the Vermont State Hospital, which also heats by wood. The type of wood-burning industrial furnace used at the hospital emits four times as many particles as an oil burner would, in spite of mechanical controls that filter out most of the particles so that they do not reach the air. Pollution from the burning of wood is currently a subject of study by the state of Vermont's Agency of Environmental Conservation.

Other states are showing concern about pollution from the burning of wood, too. For example, some towns are asked to stop burning wood on days when the air pollution reaches a certain level. More than one wood furnace, coal furnace, or fireplace per home is forbidden in new houses being built in Vail, Colorado, but small wood stoves are not counted.

In areas such as Vermont, where little natural gas is available, oil is expensive, and wood is easily available, wood stoves will continue to be popular. In valley towns where the early morning fog traps the pollution so it cannot drift away, much of that pollution will have come from the burning of wood.

Still, wood as a source of fuel for the heating of

homes and factories, and for the production of electricity, is one of the here-and-now solutions whose cost and risks are both known and more tolerable than those of many other renewable resources.

7
ENERGY FROM HYDROGEN

Hydrogen, the lightest element known, is a gas at room temperatures. It is found only in minute quantities in free form. However, hydrogen is found in water and in various fossil fuels to such an extent that those who talk about a "hydrogen economy" claim that the world need not run out of fuel. Unfortunately, at the present time it takes more energy to separate hydrogen from its compounds than the hydrogen produces as a fuel.

Many scientists are captivated by the idea of producing hydrogen from water cheaply enough to use it as a fuel. If they could, it would open the way to a nearly limitless supply of a fuel that is nonpolluting. When hydrogen burns, it combines with oxygen to produce water. There is no carbon dioxide to pollute the atmosphere, nor are there many other pollutants, such as sulfur compounds that come from fossil fuels.

Although scientists are working from a number

of approaches to split water into hydrogen and oxygen, much work remains before this can be done profitably. If the sun's energy can be trapped and directed selectively so that it breaks the chemical bonds in water, the process may be done in one step. Another approach uses two steps. Solar cells convert the energy of the sun to an electric current which is passed through special metallic materials that help in the splitting of water. In either case, the materials would be immersed in water where the reaction would take place, and the hydrogen would bubble out and be captured as a fuel.

In addition to exploring the possibility of obtaining hydrogen by splitting water, scientists are looking into the possibility of redirecting the photosynthesis of some green plants so that they would produce oxygen and hydrogen rather than oxygen, sugar, and other nutrients.

At present, hydrogen is produced commercially from natural gas for use in the manufacture of fertilizers and other chemicals. Currently, its use is limited by both cost and safety. Since it is explosive, it must be handled with great care. There are a variety of ways of storing hydrogen: as a gas under pressure, as a cold liquid under pressure (like today's liquid natural gas [LNG]), and in solid compounds of metals and hydrogen known as hydrides.

Hydrogen has already been used extensively in

space programs as a rocket fuel, but what is its future role? Even though transporting and storing super-cold hydrogen liquid presents problems that are difficult and costly to solve, some aviation experts think that it will make a good aircraft fuel. The hydrogen-fueled airplane may be the biggest single step in aircraft efficiency that has ever been taken. Certainly, the lighter weight of the fuel compared with the usual hydrocarbon jet fuel would make it possible to carry heavier loads by cargo planes.

Hydrogen, in the form of a hydride, is high on the list of candidates for fuel to power motor vehicles when gasoline and other petroleum products become too expensive to use. Experimental buses in both the United States and West Germany have already been powered by metal hydrides. These compounds release hydrogen when heated. A fleet of twenty vehicles to be serviced by a central garage is being planned as a demonstration project in West Germany. Since hydrogen produces very little pollution when it burns, the vehicles will serve well in industrial areas and crowded cities. The complete substitution of hydrogen for gasoline, however, is a long way off because of the expense of producing and distributing it. Some future vehicles may run on alternative fuels, such as hydrogen alone, while still others may burn hydrogen and gasoline.

In California, the owners of a 1979 sedan drive their turbocharged V-6 vehicle into a refueling station. The attendant quickly mates two fittings and one electrical connection. A partially automated system then checks the car's refueling system for leaks, removes air from the lines, and fills the car tank from a storage tank at the station. The lines are disconnected and the owners drive away.

This scenario sounds standard, but the fuel is liquid hydrogen and the time is now. The car is a demonstration model belonging to the University of California, where researchers are assessing hydrogen as a fuel. Even though liquid hydrogen may not be economical as a fuel for a generation, these researchers hope to help pave the way for its future use.

Many energy experts claim that although hydrogen fuel has different safety problems from those of gasoline and natural gas, they are not necessarily worse. While some people believe that hydrogen will play a small part in the energy picture of the future, others are convinced that it will be a dominant fuel.

8
COAL AND SYNTHETIC FUEL FROM COAL

If Americans live on such enormous coal deposits that their coal could supply twice the energy in all the Middle East's oil reserves, why is there an energy problem? About 90 percent of the total energy reserves in the United States consist of coal. According to some reports, it is not only the fossil fuel of the past, it is the fossil fuel of the future. Even though coal is plentiful, however, it accounted for only 18 percent of energy production in April of 1977, when President Carter unveiled his National Energy Plan. Certainly coal was king before the era of cheap oil and gas. According to some experts, coal production is expected to double by 1990. According to others, the bright hope for coal is in the production of synthetic fuels from coal.

Coal is often called "buried sunshine," for in the true sense it is a form of solar energy. Coal contains energy from the sun that fell on plants millions of years ago, when dense forests and swamps covered much of the earth's surface. Plants that died and decayed slowly formed a surface layer of peat. As many years passed, this layer of peat was covered by land and oceans. Under increased pressure and heat, this material gradually became coal.

Not all coal is alike. In general, where there was a larger amount of heat and pressure, coal with a higher energy content was produced. This fuel is found at varying depths and in different thicknesses. In most underground mines, seams of coal have been discovered that are between two and a half and eight feet deep, while in some surface mines in the western part of the United States there is coal for a depth of 100 feet or more. In the United States, it is usually grouped as eastern or western coal. While eastern coals generally have the highest heat content, they also tend to have a higher amount of sulfur than the coals with lower heat values that are found in the West.

Suppose you have a small wood and coal stove in the fireplace in your living room to supplement your usual supply of heat. Someone offers to sell you a ton of coal for $100, and someone else offers to sell you a ton of coal for $50. Which would you buy? The better buy would depend on the kind of coal, and there is a wide range from anthracite, the hardest and the kind that produces the most heat per ton, to lignite, which produces the least.

Anthracite makes up only about 2 percent of the United States coal reserve, and this is found mostly in eastern Pennsylvania. About half of the reserves are bituminous or medium soft coal, and about 40 percent of this is obtained by surface-mining areas east of the Mississippi River. Most subbituminous coal, that with a lower heating value, is found farther west. Lignite, the type that has the lowest heating value of the four major types of coal, is found mostly in Montana, North Dakota, Texas, and Arkansas.

Why not rush to coal? Some plants that switched to oil and gas when it became cheap after World War II have already converted back to coal. Many new plants are being built to use coal. But while coal is abundant, it is not without its problems.

More than half of America's coal is surface-mined, a process that is more economical than underground mining. Almost all the coal in a seam just below the surface can be removed, but surface mining, or strip mining, is controversial. Surface vegetation and topsoil are removed and in the early days of this kind of mining, ugly scars were left on the land. Today, laws protect the land against some of the damage from strip mining, but restoring it to its former condition is seldom possible.

In 1977 Congress passed the Surface Mining Control and Reclamation Act, but some challenges to the law by coal companies have created a sort of battle zone between them and environmentalists. If land cannot be restored, should it be reclaimed or rehabilitated? Reclamation means bringing the land back to approximately its former community of plants and animals with nearly its original contours and the replacement of all its topsoil. Rehabilitation means bringing the land back to a condition of stability and productivity in harmony with its surroundings. When coal is removed by strip mining, coal companies must contour the land and reseed the topsoil. Even though reclamation plans may be made before mining begins, this is often a subject of controversy in courts where portions of the law have been challenged. In all cases, it adds to the cost of the coal. Many of the people who live near the lands where strip mining is an important source of coal question the quality of the reclamation and are concerned about the need to fight for continued reinforcement of the law.

The problems of underground mining are probably better known. From the picks that individual miners use to the continuous mining machines that claw coal from seams hundreds of feet below the surface, coal mining is a complicated and often hazardous procedure. About two hundred more coal mines will be needed if coal production is to be doubled in the next ten years. Several thousand more underground mining machines, and several hundred thousand trained and experienced miners must be added to the work force. Also, large numbers of freight cars will be needed, and railroads will have to be upgraded to haul the coal from mines to users. Certainly it is more than a question of digging deeper and faster, as some people think.

Even when the coal has been taken out of the ground, transportation creates more problems. Transportation by railroads is the most common answer to moving coal, but in some cases the problems build on problems. For example, the Sierra Club forced a thorough study of a proposed train route that would carry coal from huge reserves in Wyoming, by which it was discovered that as many as forty-eight trains (of about a hundred cars each) would pass through several small towns daily. Bypassing the towns would mean building bridges, underpasses, or other expensive routes. This is but one example of the problems of transportation for the coal industry.

Coal's effect on the air has long been a controversial issue. Coal is dirty to burn, and materials pouring out of the smokestacks have effects near and far. Much work has been done to reduce the amount of harmful gases that are produced when coal burns. Some coal contains more sulfur than others, and the term "low-sulfur coal" has become familiar. How to remove the sulfur and nitrogen and other undesirable chemicals from coal has been studied extensively.

In many cases, scrubbers in the exhaust stacks remove or deplete these substances after the coal has been burned. Sometimes before it is burned the coal is "cleaned" by chemicals such as limestone and cement. Coal is sometimes burned in a "fluid-bed technique," in which a gas such as air is forced up through a bed of fine coal. This causes the coal to rise and become suspended in a virtually fluid state. This technique can be coupled with methods of controlling sulfur oxides and nitrogen oxides, eliminating the scrubbing of exhaust gases or the cleaning of the coal before burning.

When even small amounts of sulfur and nitrogen oxides spew from factory and power plant chimneys and combine with droplets in the clouds, the rain that falls is acidic; it is commonly called "acid rain." This rain falls on lakes and forests that may be hundreds of miles away from where the coal was burned. Over a period of time, fish in the lakes and vegetation are harmed, and even buildings are damaged. The gleaming white marble of the Taj Mahal, which has withstood more than three hundred years of hot sun and monsoon rains in India, may be in danger of discoloration and pitting from such industrial pollution. Ash, as well as the unwanted gases, adds to the problem of pollution wherever coal is burned. These various bits of ash, known as "particulates," add to the controversy between industry and environmentalists.

In addition to the problem of air pollution, the greenhouse effect is a concern of many scientists. All fossil fuels, and especially coal, produce large amounts of carbon dioxide gas when they burn. At the present time, carbon dioxide is being released faster than green plants on land and in the oceans can absorb it. Carbon dioxide in the air acts much like the glass of a greenhouse, allowing visible sunlight to pass, but absorbing infrared radiation from the earth's surface, which has been heated by the sun. It traps energy that would normally escape. Increased amounts of carbon dioxide mean increased amounts of heat trapped closer to the surface of the earth, theoretically causing a warming of the earth. According to a report that was made for the White House by a panel of scientists convened by the National Academy of Sciences, the amount of carbon dioxide in the atmosphere would double by the year 2030 if the present annual increase of 4 percent in the burning of fossil fuels continues. The warming would benefit some regions but would be very harmful to many areas where food production is especially vulnerable to drought and other climate changes. Since an increased use of coal is being encouraged, many environmentalists are greatly concerned about the greenhouse effect.

There are only two possible solutions to the problem of excess carbon dioxide. One is increasing the biotic sink (the green plants that absorb the carbon dioxide) by planting huge numbers of trees, and the other is by reducing the amount of carbon dioxide emitted into the air at the coal-fired plants. Both of these solutions are expensive and long-range.

Not all scientists agree with the predictions of the warming from a greenhouse effect. They believe there will be the opposite effect, because the carbon dioxide absorbs not only infrared radiation escaping from the earth but that from the sun as well. This changes the way snow recrystallizes, making it less reflective and causing the earth to cool. Sunspot activity may also counter the effect of the warming that carbon dioxide may cause. Much remains to be learned about this phenomenon.

Today coal provides about 20 percent of all the energy America uses, and about half of the electricity in the United States is produced by coal-burning plants. Since coal is being called upon to play a more important role in the future, problems that are political, social, economic, and environmental must be solved.

How will the new technologies that make synthetic fuels from coal help to solve some of the problems? No one can answer this question fully, but a look at how such fuels are made from coal helps one to understand the controversies about them.

Gas from coal is not new. Coal gasification has been carried out on a small scale here and there for a number of years, but present conditions make it a technique whose time has come. High-sulfur coal is turned into a gas that can be used in conventional steam boilers. The undesirable sulfur is separated during the gasification process and is easy to remove and handle. According to those who favor rapid expansion of the coal gasification industry, commercial gasification plants would emit only 1/5 to 1/10 the air pollutants of equivalent coal-electric plants, and would generate substantially less solid waste. However, according to a fact sheet prepared for the U.S. Department of Energy, a large amount of solid waste would be generated along with the gas.

All fossil fuels are combinations of carbon and hydrogen, and they combine with oxygen as they burn. Chemically, coal is a complex substance and different specimens are composed of different materials, but all coal has carbon and hydrogen. In the case of coal, there are sixteen carbon atoms to one hydrogen atom; in the case of so-called normal heavy fuel oil, the ratio of carbon to hydrogen is six carbon to one hydrogen; for gasoline the ratio is one carbon to two or three hydrogen. When the fuel is gas, the ratio is one carbon to four hydrogen. So it is easy to see that converting solid coal to gas means adding more hydrogen.

The first step in coal gasification is converting all

the ingredients into gases so that they can combine chemically. This is done by heating the coal in a large reactor with plenty of oxygen to help increase the temperature. The extra hydrogen is introduced in the form of steam, and in the intense heat the coal, oxygen, and water combine to form a variety of gases. These different gases are separated and the valuable gases—carbon monoxide and methane (a combination of carbon and hydrogen)—are removed and put through another chemical process in order to produce more methane and drive off the water as steam. The resulting gas is high in quality and low in pollutants, even though it can be produced from high-sulfur coal. A number of different processes of coal gasification are being developed, but the basic idea is the same. The methane gas that is produced from the coal is the same chemical that is the chief component of natural gas.

Many experts consider coal gasification a good way to unlock coal's energy because the process requires less coal. Thirty percent less coal would be required for a coal-gas plant than for a coal-electric plant to produce the same amount of energy for residential uses. But there is a controversy about whether or not coal gasification will consume large amounts of water in areas where limited water supplies must be divided among agricultural, industrial, municipal, and recreational needs. Many people in the United States are looking to coal gasification as a good way to use the vast coal resources in the United States. They favor the building of commercial-scale coal gasification plants to supplement supplies from natural gas and other nonrenewable sources, but the problems of water shortages and the release of highly toxic materials must still be considered.

Another synthetic fuel from coal is liquid fuel. Coal has yielded liquid fuels for the better part of a century. Kerosene is still known as coal oil in some areas, revealing its origin. During the late stages of World War II, Germany supplied the motor fuel for its army through the liquefaction processes, and South Africa has had a commercial coal liquefaction plant in operation since the early 1950s.

Coal can be liquefied in a number of ways, but as with the production of gas from coal, hydrogen is added in all the processes. In one procedure, pure hydrogen is added (hydrogenation); in another procedure, the coal is heated in the absence of oxygen (pyrolysis); a third process is similar to the coal gasification described, but the carbon monoxide and hydrogen are combined under pressure in the presence of a catalyst to make liquid fuel. In all of these processes, environmental concerns are present. Coal liquids are known to contain significant amounts of tar acids and other chemicals that may cause cancer. Toxic substances in the coal liquids may be a factor limiting their use. Studies will be needed to pinpoint the possible problems due to toxic substances and environmental problems associated with the transportation and the use of coal liquids.

Producing these synthetic fuels from coal is just part of America's synfuel program. Other synthetic fuels are described in the chapter on oil shale and tar sands.

9

SQUEEZING FUEL FROM ROCK: TWO MORE SYNFUELS

The promise of untold energy resources from shale oil is at least six decades old, but the great cost of mining fuel from shale has been limiting. Recent developments are helping to change old dreams into reality.

American Indians have described oil shale as the "rock that burns," for they have seen fires burn on these rocks for several days after being struck by lightning. Energy experts look to this rock that burns as a source of synthetic fuel, or synfuel. Actually, shale oil is not really oil at all, and the rock that it comes from is not really shale. The material that is valuable as fuel is a rubbery solid known as "kerogen," and the rock in which it is found is "marl," a kind of limestone. Marl is laced with streaks of kerogen, which were deposited many millions of years ago when unimaginably huge quantities of vegetable matter collected at the bottom of huge freshwater lakes. The pressure and temperature here were not as extreme as in areas where coal and oil formed, but over a long period of time the kerogen became locked into the so-called shale. Now the shale oil, as it is commonly called, looks very inviting as a new source of fuel. This synthetic fuel, however, is not easily removed from the rock.

Shale oil is plentiful throughout the world. In the United States alone, about 250,000 square miles of oil-bearing shales have been identified. While geologists estimate that as much as two million tons of fuel could be recovered from oil shales in the United States, they liken it to recovering the six million tons of gold that are said to be dissolved in the oceans of the world.

A number of methods for removing kerogen from the rock have been developed. When kerogen is heated to a temperature of 482 degrees C (900 degrees F), it vaporizes and emerges from the rock as a fine mist. So if one could manage to get the rocky area hot enough, the shale oil could be collected. This sounds easy, but consider the fact that engineers are dealing with huge quantities of rock. In one process that "cooks" the rock to release the kerogen, the shale is "distilled," somewhat the way moonshiners distill corn mash for alcohol. The rock is mined and then crushed. After this it is heated so that the kerogen is released as a gas. When the gas cools to become liquid, the rock has been left behind. Perhaps you have distilled salt water by heating it, collecting the steam, and cooling it to obtain fresh water. In small distillation processes, the unwanted material is not a great problem, but think of the tons of rocks that are left after the shale oil is removed. Some of the rock can be returned to the pit from which it was taken, but not all of it will fit, because of what is known as the "popcorn effect." The rock that has been crushed and heated takes up about 30 percent more space than it did before it was mined.

In some techniques, plans call for dumping the rock in nearby canyons where it could be packed into the bottom. According to the mining companies, the dump site could be lined with a thick layer of impervious material to prevent water from percolating through the waste rock, where it would dissolve some of the chemicals there and transport them to other places where they would cause harm. Environmentalists are concerned about the possibility of such leaching of the waste rock if the impervious material does not hold back the water after all.

The amount of rock involved is hard to comprehend. One day's operations could require that 500,000 tons of shale be mined, crushed, heated, and relocated.

One method of extracting shale oil reduces the

amount of rock that must be disposed of and eliminates some of the mining, but it still has other problems. The shale is heated in place, in a process called *in situ.* A shaft is driven deep into the rock, and a large chamber is mined out of it. Then the shale above the chamber is broken into pieces by the use of explosives, and this rubble falls into the chamber. A hole is drilled into the shattered rock and air is injected. Then a fire is started at the top; it burns downward, burning some of the oil but vaporizing most of it from its heat. The vapor, or gas, is condensed by the cooler rock in the chamber, and this liquid collects at the bottom. From there it is pumped to the surface.

Environmentalists and other critics of the *in situ* process are concerned about the possible release of dangerous chemicals that could seep into the region's groundwater. The water needed for the extraction processes is another problem, since shale is found in semi-arid regions where water is already in short supply. Some people say that the shale extraction would use from one to five barrels of water for each barrel of oil, but the mining companies disagree. Experimental plants are now trying to show that shale oil extraction is economically and environmentally sound.

Tar sands are a promising source of synthetic fuels. For this "oil in place" is present in enormous quantities. The famous Athabasca tar or oil sands in Alberta, Canada, are estimated to contain almost twice the recoverable conventional oil reserves in the entire world. These deposits were laid down about 110 million years ago. Other very large deposits are found in China and Venezuela.

The tar sands consist of bitumen, a petroleum deposit, contained within a framework of quartz sand grains, with small amounts of other minerals. As in the case of oil shale, huge amounts of materials must be handled to separate the organic and burnable parts from the sands. The usual method of removal is to take away the topsoil and strip-mine the sands, then move the materials by conveyor belt to the extraction plant, where they are broken into sand and bitumen. In one operation in Canada, 250,000 tons of material are dug out in a day, and the same amount of waste material must be disposed of in some other place. Syncrude of Canada, Ltd., has an operating plant that should produce 130,000 barrels of oil per day when its maximum output is reached. By the 1990s, production from tar sands in Canada could reach one million barrels per day. This is about one-third of Canada's requirements.

The second approach to tar-sand extraction is sometimes called the "huff and puff" method. The oil in tar sands is separated from the rocky material by direct heating to about 80 degrees C (176 F), or by passing steam through the tar sands. When heated this way, the oily substance flows freely to the surface, where it can be recovered and the rocky material can be washed away from it. This process is used to get at some of the deeper tar-sand deposits. For every barrel of oil, two tons of material must be processed. When the oil is processed further, such as by increasing the hydrogen content, it can be used in place of most petroleum products.

Some energy experts consider tar sands to be one of the most promising alternative sources of fossil fuels that can be available for the near future. When it was believed that the deposits in the United States were equivalent only to one billion barrels of oil, there was not much interest in them. Now that estimates run thirty times this much, and the price of oil has risen, tar sands are more appealing as an alternative source of energy.

No matter where they are found, tar sands are an expensive source of fuel. It takes a great deal of energy to extract the oil and a great deal of money to buy the equipment. For example, at one plant, a fleet of trucks that use large amounts of gasoline criss-cross the ground hour after hour, day after day, to carry away ton after ton of waste. For every 50,000 barrels of oil produced, about 100,000 tons of tar sands must be mined and disposed of. Huge crane-operated buckets pick up 130 tons at a bite, but they require large amounts of energy for their operation. Whirling bucket wheels transfer tar sands to conveyor belts which carry almost 12,000 tons an hour to the separation plant. All of these processes require energy. It takes a half barrel of oil to produce a barrel of synthetic crude. Those who favor developing the synthetic fuel industry quickly argue that even though the net energy gain may be only 50 percent, even this amount of oil will help fill our energy needs.

While some energy experts support a crash synfuel program, others say rushing into such a program may mean repenting it later. Some of the problems have already been noted.

Oil shale, tar sands, and coal are all sources of synthetic fuels. Opponents of the rapid development of a synfuels program point out that it would require both strip mining of coal and the mining of oil shale on a scale so huge that hundreds or even thousands of square miles of land would be involved. In addition to tearing up huge chunks of territory, large-scale programs would create the need for the establishment of new communities. The problem of the possible leaching from waste materials has already been mentioned. Another concern is the great quantities of water needed in mining and processing raw materials, water that might be taken from agricultural lands.

In June of 1980, a new bill that is a major step in

the United States Energy Program was signed into law. Synthetic fuels are a high priority of the legislation, which calls for the development of alternate fuel sources. The short-range goal is to provide the financial and legal assistance to enable private industry to build ten synthetic fuel plants by 1987. It is hoped that they will produce 500,000 barrels a day, the equivalent of eight percent of our oil imports in 1980. The long-range goal is two million barrels of synthetic fuel a day, or roughly one-third of the imports for 1980.

10
ENERGY FROM BIOGAS AND TRASH

In China a family with two pigs produces enough waste fuel for a family to do all its cooking with it and light one lamp. The waste, including that from pigs and humans, plus a few leaves, produces methane gas in an inexpensive biogas unit, one of the units that is used by seven million families in China.

Biogas is any gas that is produced by the decomposition of animal and/or plant wastes in a limited amount of oxygen. Bacteria that thrive in conditions under which oxygen is limited are called "anaerobic," so the devices are called anaerobic converters or biogas converters. Since the gas that is produced is methane, they are also called methane converters. The units are also known as "digesters" and as "generators."

In addition to the millions of converters in

China, there are many more millions in use in other countries, such as India and Korea, and the number of them is expected to grow rapidly. For example, in India there are plans to build 18.75 million family-size biogas units and 560,000 community-size plants. If these plans are carried out, biogas may supply about 44 percent of the country's needs by the year 1990.

The year 1990 is a target date for self-sufficiency in American agriculture, and biogas is expected to play a major part in this program. While American farms now produce three times as much as they did about forty years ago, they consume much more fuel for farm machinery. Increased amounts of energy are used in the production of fertilizer and for other needs. In fact, agriculture was the nation's largest consumer of petroleum in recent years. In one year, energy for food production accounted for 40 percent of the total United States energy consumption. But a wide-scale effort is now being made by farmers and ranchers to substitute renewable forms of energy for scarce fossil fuels and to conserve energy in every possible way.

Here and there in the United States, farmers have been converting some of the animal and plant wastes to fuel for heating homes, poultry and livestock houses, and for drying grain. Methane gas produced by the biogas generators is not a new form of energy for many farmers, but the increase in the amount of methane that is made this way is new.

Use of biogas holds great promise for a large-scale reduction in the amount of petroleum needed to supply the United States with food. Cooperative extension offices in more than three thousand counties throughout the United States are encouraging farmers to become self-sufficient in their energy needs by using biomass in the form of wastes from plants and animals to produce methane. The field residue from grains and other plants, cottonseed hulls, and byproducts from fruit- and wood-processing operations that were once wasted are now being used to produce energy on farms. Manure from poultry and livestock adds to the increasing sources of biogas. The total value of collectible organic wastes has been compared with 3 percent of the oil demand and 6 percent of the natural gas demand used by the United States in a single year, according to an estimate by the Bureau of Mines.

Although some of the organic matter used for the production of biogas is trash, not all of this kind of material has been wasted in the past. It had value as fertilizer when it was returned to the soil. Farmers must consider whether there is greater value in using organic matter to produce energy or to replenish the soil.

Where the trash is a nuisance, energy obtained

from it seems less important than getting rid of the trash. For example, getting rid of chicken manure is a serious problem in Maine because chemicals from the manure pollute the water supply. Here, some biogas converters are helping to relieve this problem, while providing energy in the form of methane.

The methane that comes from the decomposition of manure and plant wastes in a limited amount of oxygen is produced naturally as well as in the converters mentioned above. Methane is familiar to some people as marsh gas, for it is commonly produced in marshes. It is also a product of sewage sludge, and it is from sewage that some of the gas is channeled into energy production.

In Riverton, New Jersey, the methane gas that was naturally generating from the garbage in the town dump was put to work at the beginning of August 1979. For several years, the methane from this landfill had been considered a hazard because it was spreading to nearby farms and threatening crops. But the owner of the landfill site, along with the local utility company and a corporation that operates a powdered-metal factory, cooperated in developing the landfill for fuel production. Now the methane gas generated at the dump provides a portion of the company's energy requirement. The supply of gas from the site is expected to last for ten years.

Methane is helping to supply needed energy in a

variety of places. Some landfills contain billions of tons of material capable of producing gas that can be used profitably by urban utility companies. This method of obtaining energy becomes more attractive as the price of oil soars.

When one considers the energy potential of all kinds of trash that is discarded by a whole city, the picture looks even brighter for this kind of energy production. Each year about two hundred million tons of municipal solid waste must be disposed of in the United States. Think of all the things that go into the trash in the city. Metal cans, plastic toys, old tires, orange peels, razor blades, diapers, lawn sweepings, boxes, bottles, refrigerators . . . An almost endless list of things make up the refuse that has to be disposed of in some way. Combustible paper, wood, some kinds of plastics, and other materials that can be burned make up 70 percent of this waste.

Such city wastes are now producing energy in a growing number of places in several different ways. In some systems the trash is sorted and the burnable material is compacted into pellets that can be used along with coal in electric generating plants. The pellets are called "refuse-derived fuel." They do help to stretch the coal or other fuel that they are combined with, but they cannot be used alone because chemicals in the pellets can cause rapid corrosion.

A resource recovery plant in Milwaukee, Wisconsin, makes refuse-derived fuel pellets and recovers 90 percent of the city's waste for new uses. In one year, resources that were previously dumped on the land produce about 150,000 tons of solid fuel for electric power generation, 15,000 tons of ferrous metals, 1000 tons of aluminum, 10,000 tons of newspaper and corrugated paper, and 20,000 tons of a material known as "glassy aggregate."

If you live in Milwaukee, you try to keep your solid trash dry to help make the system work more efficiently. Imagine a day in which 1500 tons of refuse arrives at a plant for conversion into energy and scrap. All kinds of things are mixed together with the exception of the newspapers, which people have bundled together so that they can be easily removed from the other trash.

The city trash is dumped by trucks onto the floor of the "tipping room," where large vehicles push it into a conveyor trench. The material is inspected by workers who remove clean corrugated paper and packages of newspapers so they can be recycled as usable paper. Then a huge shredder breaks up the trash and the garbage into small pieces. Almost anything can be shredded by this giant machine. From here the material travels on a conveyor where it is automatically sorted. The light material that can be burned is removed from the heavy material that falls to the bottom of a three-story-high device. The light material is shredded a second time, and then it is compacted directly into trucks that carry it to the electric power generating plant.

Heavy trash, which is now on the lower level, is moved on the conveyors to a series of magnets that pull out the materials containing iron and steel so that these can be carted away for recycling. The rest of the refuse is sent through a vibrating screen or a rotating drum where glassy aggregate is separated. Metals that were not removed by the magnets are separated from the rest of the remaining trash, and what is left is sent to landfill. The bulk is only 10 percent of what it was originally, so the problem of space for trash is greatly reduced.

The Milwaukee resource recovery plant is just one example of a growing number of recovery systems. Trash at Saugus, Massachusetts, is converted to energy by a different process. Here heat from the burning refuse turns water into steam. Water is circulated in steel tubes that jacket the exterior of the incinerator, forming a wall of water from which the process gets its name of "water-wall incineration." There are well over a hundred plants of similar design throughout the world.

As much as 1500 tons of solid waste from the Boston North Shore Communities is disposed of each day, producing steam for the nearby General Electric plant. This means a savings of fuel oil as well as space that would be used as a dump.

A third method of recovering energy from trash in large plants is known as "pyrolysis." Wastes are converted into fuels by "baking" them in an oxygen-starved environment. The typical pyrolysis plant uses coarse shredded waste material in a hearth where temperatures are extremely high. The resulting gas is largely carbon monoxide, a gas that can be used by industrial customers but is not suitable for heating homes because it is lethal. Pyrolysis gas can be sold for conversion to other chemicals, one of which is methane. Oil from the process has value as fuel, but it requires special handling. Molten slag, which is left by the noncombustible materials, flows into a quench tank and is removed for disposal.

A number of other new processes for converting trash into energy are being explored. One powders solid waste into a fuel that can be used along with coal. Although this holds promise, there is danger of explosion because of the dustlike composition, and special handling is necessary.

Obviously, changing trash to energy and recovering other resources is not a simple procedure. There are many problems that must be solved before a plant can operate economically and stay within environmental regulations. If such a plant were to be built near your home, would you question whether or not odors might fill the air and be carried by the wind to your nostrils?

Controversy over plans for the Arthur Kill power plant on Staten Island in New York City ranged from charges that the water used in the process could transmit a wide variety of diseases over four of the city's five boroughs to expert statements that there would not be any adverse health effects.

Although some trash-fired plants are beset with problems, such as spreading nauseating gases over nearby neighborhoods and exposing workers to unsafe levels of dust and germs, others have been cited for environmental excellence. As problems are solved and as the cost of fuel continues to rise, resource recovery projects become more attractive in the United States. Early in 1980, some 430 facilities were engaged in converting municipal, commercial, and light industrial waste into useful energy at various places around the world. Only seven were in operation in the United States at that time. While some experts lament the slow development of such plants, others express great concern about their effect on the environment.

If all the municipal solid waste that is produced in the United States could be collected and processed, it might generate about 5 to 6 percent of the total energy requirement of our electric utilities. But before energy from trash is used on a large scale, hazardous and undesirable chemicals must be prevented from escaping into the atmosphere.

From the family with two pigs that use their

wastes to produce methane in a small converter to the large plants that supply part of the fuel needed to produce electricity for the people who live in large cities, trash contributes to the energy supply. Its use is a topic of growing interest and controversy.

11
ENERGY FROM WAVES, TIDES, AND OCEAN HEAT

Solar sea power is an exciting new approach to the search for new sources of energy. The tropical oceans collect and store vast amounts of solar energy, and the waters move constantly as a result of changes in the weather and the pull of the sun and moon. Three major sources of energy from the sea are wave power, tidal power, and power derived from the differences in water temperature. The last approach is known as ocean thermal energy conversion (OTEC).

Consider the waves that pound against the coastlines and roll onto the beaches hour after hour, day after day. Anyone who has seen the destruction of beaches and buildings along the coast is aware of the energy in waves. Scientists have calculated that the amount of energy in the world's ocean waves is greater than that in the world's oil supply. The energy in ocean waves is truly awesome. How can some of this energy be harnessed?

Although obtaining electricity from the energy of the waves may sound impossible, models that demonstrate how this can be done are already at work. One model is the Salter duck, a device named for its inventor and the way it bobs up and down on the water. This pumping machine is driven up and down by wave action. It is hoped that a number of these modules may be connected with an underwater crankshaft that will be turned to produce electricity. Some day this device and/or similar models may be placed several miles off the coast. Models more than a mile wide may tap vast amounts of the energy from waves and send it by underwater cables to electricity grids on the shore.

A full-scale Salter duck would have little resemblance to the animal for which it was named. Estimates indicate that almost five million pounds of concrete would be needed to construct an ocean-wave power plant that would produce as much electricity as a conventional plant. Each unit would be about ten by thirty feet in size and would transmit energy to a generator as it rocked back and forth in the water. Units connected side by side might stretch for several miles, sending power to the shore through cables spaced a mile apart.

Another device working in a similar manner is known as a raft. In this, a series of floating pontoons would produce the motion to drive water through turbines.

A scale model of a different type of system for extracting energy from ocean waves has been constructed by the Lockheed Missiles and Space Company, Inc. The test model, known as the DAM-ATOLL, was built to a scale of 1/100.

A DAM-ATOLL is an artificial atoll in which the shape has been mathematically refined to make waves spiral into the core of the unit, producing a whirlpool action. If you stir a bucket of paint with a stick you produce this kind of action. In the DAM-ATOLL, a large cylinder is placed over a central core which is continuously filled with water. As the waves travel over the dome, their speed slows as the water depth decreases, causing them to bend. The waves spiral into the center of the dome, and through a carefully devised structure they are introduced into the core in such a way that a fluid flywheel is created. A turbine wheel is located at the bottom of the central core and the wheel turns, providing energy to drive an electrical generator. Thus usable energy can be obtained from the pulsations of waves.

Full-scale models will be eighty meters in diameter and they will be expensive. As with other solar systems, the initial cost is high compared to that of a fossil fuel plant, but free solar fuel (in this case in the form of wave activity) should produce an overall savings in the long term.

Most research in obtaining usable energy from waves is quite new, but the Japanese have already made practical use of this kind of energy. They use the up-and-down motion of waves to compress air and drive a small generator in a lighthouse. Research on wave energy is going on at present in Sweden, Great Britain, and other countries.

The use of tidal power is not new, but it is limited to those places where the difference in elevation between low and high tide justifies the construction of tidal plants. The world's largest tides occur in the Bay of Fundy, which separates New Brunswick and Nova Scotia in eastern Canada. In the United States, the only possible sites for practical power plants are Passamaquoddy Bay in eastern Maine and Puget Sound in western Washington, bordering Seattle, and the south coast of Alaska. Unfortunately, some of these areas are far away from places where large amounts of power are needed. A tidal plant operates at Kislaya Guba in the Soviet Union, and one produces electric power at LaRance, France. These plants operate much the same way as hydroelectric dams. When a dam is constructed across the inlet, the tide that rushes in turns the turbines in the dam. These turn the generators to produce electricity. When the tide goes out, the water runs through the turbines from the reservoir and electricity is produced once more. Unfortunately, tidal power is not constant, but it *is* renewable. It is nonpolluting and has minimal environmental side effects. Changes in the action of the tides due to power projects affect the sedimentation patterns and may have some effect on the fish and mudbank flora and fauna of a region, but these changes are not expected to be severe.

Since energy from moving ocean water is free and will never be depleted, tidal power has the advantage of a free and inexhaustible fuel supply. While hydropower is affected by droughts, the tides move in unchanging cycles.

Although tidal power is not expected to contribute as much to the energy picture as other forms of solar energy such as wind, hydropower, or even energy from waves and differences in ocean temperature, tidal power conversion could become a substantial and economic supplement to the other forms of renewable energy.

For a hundred years, scientists have considered using the temperature difference between the warm ocean surface and the cold ocean depths as a means of obtaining energy by the method known as ocean thermal energy conversion (OTEC).

While tropical waters soak up energy from the sun, the polar ice caps are melted by it. Deep, cold currents flow from the polar regions toward the equator. There is a band of water ten degrees north and ten degrees south of the equator that is about 26.66 degrees C (80 degrees F) on the surface, and about 4.44 degrees C (40 degrees F) at a depth of approximately three thousand feet. If a fluid, such as ammonia, is piped through the water at the surface, it comes to a boil and turns into gas. When this gas is piped through the cold waters, it changes back into liquid form.

When the liquid ammonia passes through a heat exchanger, warm ocean water changes it into a gas under pressure. This gas operates the turbine on a specially built ship or a floating plant, and electricity is generated. The gas expands, is cooled in another heat exchanger, and becomes liquid in the cooler waters. The cycle is repeated again and again.

The process of ocean thermal energy conversion (OTEC) was demonstrated on shore and on a ship off the coast of Cuba in the 1930s, but a hurricane destroyed the cold water pipe and the project ended because of a lack of money. In 1979 the world's first "at sea" ocean thermal plant accomplished its mission of demonstrating the technical feasibility of OTEC. This mini-OTEC operated for several months off the Kona coast of the Island of Hawaii, producing electricity and providing information about possible problems of a plant operating in seawater.

Mini-OTEC gave an indication that gas that might have been trapped in the cold water, and debris that might have clogged the inlet screens for warm and cold seawater, were not the problems they were considered to be. Biofouling, the deposit of microscopic-sized plants and animals as slime on the seawater side of heat exchangers, could have reduced the amount of heat used to vaporize the ammonia, but researchers appear to have eliminated this problem by injecting a minute amount of chlorine into the seawater. Only one-tenth of one part of chlorine per million parts of water was continuously injected into the seawater in the heat exchangers.

In any ocean thermal energy conservation system, some electricity is required to run the pumps, the compressor, navigational lights, and the refrigerator. On the small plant off the Kona coast, this amount of electricity was a high percentage of the amount produced, but in a large plant the percentage needed is expected to be only about 30 percent of that which is produced. This is much greater than the power taken by conventional plants for their internal needs, but in OTEC the fuel is free.

An OTEC plant that would provide enough electricity to fill the needs of 200,000 people would be 250 feet in diameter and 1600 feet long. It would weigh 300,000 tons. The structure would be a floating platform with crew quarters and maintenance facilities, and with turbine generators and pumps attached around the outside.

Since the heat difference between surface water and deep water is not large, huge amounts of water will have to flow through the heat exchangers of OTEC plants. The amount of water has been compared to the average flow of the Missouri River at Omaha, which is about 50,000 cubic feet per second. Materials and design for a large ocean device present an engineering challenge, but it is one of scale rather than technology. There is agreement that OTEC plants have to be massive to be economically profitable. Construction materials are available and construction methods are known. Since this method is environmentally benign, and its use of solar energy is not affected by weather or the day-and-night cycle, many people find it a very attractive path to pursue.

Opinions on the promise of ocean thermal energy conversion vary greatly. Some experts call it the biggest gamble in solar power, while others believe that massive OTEC plants may be adding to the energy supply by turning solar heat to usable electricity by 1985.

12
CAPTURING STAR POWER: CONTROLLED FUSION

Why not copy the sun's own method of producing energy to add to the earth's supply of usable fuels? In the sun and other stars, countless numbers of hydrogen atoms are changing to become helium atoms at all times. As they do, a small fraction of the mass of the hydrogen atoms changes to energy. This happens in the sun on a tremendous scale, with the conversion of ten billion pounds to energy every second. The earth soaks up only a small part of the energy emitted by the sun, just one two-billionth. Even this fraction is an amount of energy too vast for human understanding.

The process by which the sun's energy is apparently produced is known as "fusion." This is a nuclear reaction that is, in a sense, the opposite of fission. One might think of the earth as being supplied with energy from a huge nuclear reactor that is a safe ninety-three million miles away.

An example of the release of the energy of fusion on earth is the explosion of a hydrogen bomb. Can such energy be harnessed? The problem of releasing fusion energy in a controlled manner is the subject of vast research. Many experts are at work trying to apply the heat of fusion to the firing of boilers that produce steam.

Fuel for fusion is plentiful. Researchers work with deuterium, a form of hydrogen, and with tritium, another form of hydrogen, since these offer the most promise.

Deuterium is found in water. Only a pound of it occurs in about 60,000 pounds of water, but there is so much water on the earth that there are probably ten trillion tons of deuterium. Since the energy in one ton of it could produce sixty million times more energy than a ton of coal, you can see that there is plenty of energy potential here. The United States Department of Energy reports that the estimated supply of energy from deuterium could last ten billion years at current world consumption rates, a period that is many years longer than the life of the earth itself thus far. Certainly there is no danger of running out of this kind of fuel.

The tritium that is used in fusion research is made from the metal lithium. Some scientists believe that the supply of lithium is great enough to last for thousands of years as a fuel for fusion reactions. It is widely distributed in nature: it is found in plants, animals, soil, and ores. Some people believe that the economical supplies are limited unless lithium can be extracted from seawater at a very low cost. However, the supply of lithium is not really a serious problem, since fusion reactions may eventually use only deuterium.

Fusion holds the promise of bringing abundant energy to the farthest reaches of the earth. The dream of harnessing nuclear fusion reactions of the kind that power the sun to produce unlimited energy is not new. Since 1952, when the hydrogen bomb released uncontrolled fusion energy, scientists have looked to the day when they could harness this kind of energy. Hopes have risen and faded again and again through many years of research. Since the fuel is plentiful, what is the problem?

In the fusion process, hydrogen nuclei combine to create helium and release one particle, a neutron, which flies off with tremendous energy. Fusing together the nuclei of billions of light atoms, such as ordinary hydrogen, or the heavier forms, deuterium and tritium, to produce usable amounts of electricity is a tremendous task. The nuclei have a positive charge, and they repel each other. The forces that bind together the constituents of the nuclei of atoms must be broken before energy is released. The combination, or fusion, of the small nuclei of two atoms to form one larger atom and to release energy takes a great energizing force to get going. As in the case of fission (splitting the nuclei of atoms), once the reaction is under way, energy is released in the form of heat from the unbelievably rapid motion of the nuclear particles that are freed in the reaction. In the case of the hydrogen bomb, an atomic fission bomb was used as an energizing force. In the case of the sun, the force is solar gravity.

The earth's gravity is very weak compared with that of the sun. It takes tremendous speeds and temperatures of nearly 100 million degrees Celsius to smash atoms together so they fuse their nuclei and release energy. Obviously, a controlled fusion reaction is extremely difficult to achieve. Exact basic requirements must be met for the proper density as well as for the necessary time and temperature if nuclei are to overcome their repulsions. Any one of these factors (speed, temperature, density) is easy enough to regulate, and scientists can even arrange two of them, but they need to put all three of them together for controlled fusion. According to the British physicist J. D. Lawson, the proper material must be confined for one second at 100 million degrees at a density of 1000 trillion particles per cubic centimeter. (This sounds like a large number of particles, but actually this density is far less than that of air.)

How can any material be confined at such a high temperature? The sun uses its gravity, but this cannot be mimicked on earth, where, as mentioned earlier, the gravity is much, much weaker. The hotter atoms are, the faster they move, so scientists knew early in their research that the nuclei of atoms must be heated to extremely high temperatures to be able to fuse. Atoms must be made to move unbelievably fast to overcome the way they naturally repulse each other and to get them close enough to each other to make fusions likely.

At very high temperatures the electrons that normally spin around the nuclei come loose and separate from the nuclei. The material takes on a special form known as "plasma." Scientists call solids, liquids, and gases the three states of matter. Plasma is known as the fourth. They have studied this fourth state intensely for twenty-five years in the science known as plasma physics, the field of research that deals with the behavior of matter in this state.

It is obvious that any ordinary container would vaporize at the high temperatures needed for fusion of the nuclei in the plasma. One popular and promising approach is the use of a doughnut-shaped device called a "tokamak." At Princeton University, scientists have been heating plasma at temperatures up to seventy-five million degrees in a tokamak in which magnets surrounding the doughnut keep the electrified plasma away from the walls. In this and all tokamaks the plasma particles carry an electrostatic charge so the direction of their movement can be influenced magnetically, somewhat the way a magnet acts on iron. The area between the plasma and the container is a vacuum, a sort of nonmaterial liner, which prevents particles from touching the chamber. In a tokamak, a strong electric current is induced in the plasma itself and combined with the action of the strong magnet around the vacuum chamber. These forces give the total magnetic field a spiral form, and have the advantage of heating the plasma with the current and helping to stabilize the plasma at the same time.

In another controlled fusion technique, plasma is created and confined in a straight chamber, in which the plasma is made to bounce back and forth in the "mirror machine." As the charged particles move toward the ends, where the magnetic field is the highest, they are reflected back toward the middle. The particles may be reflected back and forth many times before they escape.

A newer and entirely different approach to controlled fusion uses lasers. Researchers blast a laser beam at a minute, frozen pellet of deuterium and tritium, heating the outer layer of the pellet and causing it to collapse inward. This increases its density and temperature, and fusion occurs.

It was hoped that laser bombardment of pellets would release a great deal of energy. However, when the powerful laser system at the University of California's Lawrence Livermore Laboratory began its operation, the lasers delivered less energy than expected. Although hopes have now been scaled down, there is still the possibility that some form of pellet-crushing may prove to be economical early in the next century.

Fusion researchers have revived an idea that was proposed as long ago as 1964 but considered impossible then. Now they are experimenting intensely with the pellets, with electron beams and light particles, and the nuclei, or ions, of hydrogen. One machine at Sandia Laboratories in Albuquerque, New Mexico, that sends thirty-six ion beams to a single target will have an additional thirty-six beams operating by 1983. At this point, scientists hope to achieve a net energy gain.

These are just some of the approaches that are being explored in huge, expensive, experimental equipment. The Department of Energy estimates that the total funding required to arrive at a commercial demonstration plant that provides energy from controlled fusion is about fifteen to twenty billion dollars. But fusion power is regarded as a long-term energy resource that is well worth exploring.

After successful scientific experiments comes the

construction of demonstration reactors. The commercial fusion reactors come years later, but plans for Starfire, a commercial model, are already under way. The cost of fuel for fusion reactors would be very, very low, and the supply would be almost limitless.

What about pollution? Unlike the situation in fission reactions, there is no possibility of a meltdown, since any malfunction will destroy the plasma and stop the reaction. Although the radioactive wastes from fusion reactors will not be as hazardous as those from fission reactors, fusion will not be entirely free of radioactivity. Large amounts of tritium will have to be handled in deuterium-tritium reactors, but tritium emits only low-energy radiation, and its time of retention in the human body is brief. The stainless steel reactors will become radioactive and will need to be replaced, so storage of used, radioactive stainless steel must be taken into consideration.

Fusion has been called the pot of gold at the end of the rainbow, the hope for the world, for the children of tomorrow, and for generations to come. No wonder there is increasing pressure from scientists and concerned citizens to push ahead with research that may make controlled fusion an important part of the energy picture of the future.

13
CONSERVATION: THE BEST HOPE

"Conservation" has long been an unglamorous word. Today it is the bright hope of the immediate energy future and an important part of plans for the year 2000 and beyond. For most people, conservation has taken on a new meaning. Instead of cutting back and lowering the quality of life, conservation has become the way to help prevent these things.

Stop waste. Make transportation, heating systems, and other large energy consumers more efficient. Do the same for millions of little things. Add conservation to the other untapped energy resources. These are the new slogans.

More efficient use of energy has been called the easiest and the best source of energy for the future. No wonder conservation has taken on a new meaning.

What is the true potential of conservation? Is it just a promise of an easy way to get more energy so that people will not panic about gloomy days ahead? Certainly conservation is not the big technological "fix" that many people are still hoping will solve the problems. Those who realize there is no single answer are aware that one of the best ways to get more energy today is to stop wasting most of it.

Suppose your city discovered that it did not have enough fuel to provide for the needs of the winter ahead. Newspaper headlines might suggest ways to cope when the lights go out. Panic might reign. Or people might begin to take action, looking for ways to save energy. New programs would start to help citizens conserve. If people did not cooperate, there would be blackouts, shortened work weeks, loss of jobs, and other problems.

The above situation actually did happen in the City of Los Angeles as far back as November 1973, when the Department of Water and Power in Los Angeles realized that more than half its annual consumption of oil would not be forthcoming because of the Arab oil embargo. Through a program that leaned heavily on the cooperation of consumers, who were threatened with increased charges and the possible cutoff of service if consumption of electricity were not reduced, the city overreached its target. The target for reduction in the use of electricity was 12 percent and the actual reduction was 18 percent. This program affected residential, industrial, and commercial users, and the response was good in all areas. Commercial consumers showed the biggest drop in use, a reduction they accomplished through better control of lighting and air conditioning. The whole program illustrated what could be done when the need to prevent waste was really felt. Of all the paths that may be and should be pursued, energy efficiency is probably the most underrated resource.

Studies show that about 20 percent of the energy used in the United States is consumed by buildings, for heating, cooling, and lighting, and that more than 50 percent of that energy is wasted. Energy users, large and small, can make a difference here.

Many conservation-conscious people are searching for ways to reduce energy waste. Some of this effort begins at home, where houses are found to be leaking heat or cool air through numerous pores. The house has been called a leaky envelope, and the areas where leaks are most common have been well established. These include windows, doors, baseboards, wall outlets, holes where telephone wires and plumbing pipes enter the house, openings around dryer vents and exhaust pipes, and around sink and bathtub drainpipes as they exit from the house.

Low Cost, No Cost is a pamphlet distributed free by the Department of Energy that suggests ways to plug up the holes in a house and other ways to cut down on energy use at home. The energy-saving tips suggested in the pamphlet can be carried out in a short time with a few simple tools that almost everyone already owns. Some materials must be purchased in order to follow all the suggestions, but it is estimated that the energy savings as a result of this conservation will be more than realized the first year. If you wish to obtain this booklet or other information, write to the Department of Energy Technical Information Center, Box 62, Oak Ridge, TN 37830.

Perhaps an energy audit for homeowners is available in the area where you live. Your local electric utility company or state energy office might send someone to your home to point out ways in which you can conserve energy. Audits may contain as many as one hundred factors, some of which are adjusted for local weather conditions. Trained auditors who visit households to inspect conditions make a variety of energy-saving recommendations. They may measure and analyze the variables affecting energy use in your house, such as the total square footage under one roof, the number of people who live there, and the value of present insulation. They may calculate the cost of energy-saving products and determine the amount of money that would be saved over a period of time if these products were installed. Auditors are not interested in selling anything other than conservation.

Consider the energy that is used in schools in your area. Many school administrators are cooperating with those who conduct energy audits to see how much energy can be saved in school buildings. Energy was not a concern when many schools were built in the 1950s and 1960s.

Today auditors are asking the following questions: Are unused areas heated, cooled, and/or lighted needlessly? Are thermostats set back at night? Are new high-efficiency light bulbs being used? Are leaks being plugged? Could more insulation save heat? How can energy consumption be reduced in commercial office buildings and large apartment houses?

In some communities, large numbers of people have gotten together to show what a difference conservation can make. Portland, Oregon, developed the first formal "cookbook" for energy planning at the municipal level. This guide was two years in the making, and tells its story in eleven volumes with a total of 1500 pages. Large numbers of other cities are now using this guide to chart their conservation planning.

The Portland Plan is also called "the Portland retrofit story." "Retrofit" is a word borrowed from the space industry that describes the upgrading of a complex system through the use of improved parts. After citizens' committees produced the guide for making the retrofitting of weatherization (protection against weather) mandatory in all existing dwellings, the Portland City Council voted the plan into law on August 5, 1979. According to the plan, builders and homeowners would be encouraged to weatherize voluntarily during the first five years. After 1984, an owner may not transfer clear title of a structure unless certain standards are met.

The Portland Plan treats many facets of energy use besides weatherization. For example, it encourages housing close to areas where people can work and shop, provides tax and financing incentives for solar devices, recycles wastes, and provides conservation assistance to tax-free institutions.

Portland's energy use may be cut as much as 30 percent when the plan becomes fully operative. In reality, savings may be multiplied more than a hundredfold as the plan spreads to other cities.

A program for winterizing a small community was put into action in Fitchburg, Massachusetts, in 1979. Concerned residents, with the support of volunteers, help those who cannot insulate their own homes, attach flow restrictors to showers and faucets, clean the coils in the backs of refrigerators, and carry out other inexpensive energy-saving tasks. High school and college students, members of the PTA, neighborhood centers, senior citizen groups, and other community-minded organization members have joined together to make certain that all of the homes, business firms, school buildings, and city buildings in Fitchburg are caulked and sealed against needless energy waste.

In Davis, California, where bicycles are used for about 25% of all local trips, people are working hard at cutting energy consumption on all fronts. Here and in many other communities, energy conservation has become a way of life. Electricity consumers in many rural areas are showing that energy conservation works, too. They cut their normal use of power nearly in half during 1978, according to America's Rural Electric Cooperatives and Power Districts. This saving is estimated to be enough electric power to serve all the homes in Washington, D.C., for a year and a half. America's 1000 rural electric systems report that energy conservation really works.

In addition to cutting down on use of electricity and heating fuels, action in the world of transportation has included the speed limit of fifty-five miles per hour in the United States, the trend toward cars that get better fuel mileage, car-pooling and van-pooling, the increased use of public transportation, and the trend to vacationing nearer home. Publications such as *How to Save Gasoline and*

Money are available free from the Department of Energy, Box 62, Oak Ridge, TN 37830.

New technology plays a part in the conservation picture. Weight reduction in automobiles, new engine designs, and research in other areas are helping in the world of transportation. New materials are conserving energy in the building industry. For example, better windows and improved insulation are popular in new housing as well as in old. Some home buyers look to energy-efficient houses that include a solar energy system for part of their supply of heat, along with conservation features. Others depend entirely on a conservation design with thermally massive walls, windows that face south, and sophisticated thermal controls. Conservation design alone can bring down the need for heat by one third when compared with a standard house. Almost all home buyers are aware of the need for energy efficiency.

However, even conservation can have its hidden health threat. When ventilation is reduced by the plugging up of the openings of the leaking envelopes in which people live, and when conservation-minded builders seal homes against heat or cool air loss, indoor air pollution can be serious. Building and insulating materials that emit radon, a radioactive gas, and formaldehyde, a toxic chemical, should be painted with sealants. Houses should be aired briefly each day if there is no system for ventilation.

In Scandinavian countries, a system for freshening the air has been used for many years. Simple devices that use only small amounts of energy pull in fresh air and blow stagnant air out through separate ducts in a common wall. Some heat from the indoor air is absorbed by the air coming from the outside as it passes through the common wall.

More sophisticated heat-exchange systems that resemble air conditioners are available for airtight homes. If you live in a home that is sealed or in the experimental home in Mt. Airy, Maryland, you will need such a unit. The National Association of Home Builders' Research Foundation built an experimental home there that has leakproof triple-glazed windows, a weather-stripped magnetically sealed front door, and sheets of plastic in the floors, ceilings, and walls. These and other conservation efforts keep the house airtight. The builders boast that they can heat the house with a hair dryer, but they found it necessary to install a heat-exchanger to prevent indoor pollution.

Conservation-minded people are benefitting more from heat pumps than ever before. Such pumps remove heat from the interior of a building in summer and discharge it outside. They extract air from the outside in winter, warming it in somewhat the opposite of the way a refrigerator works. Then the warm air is pumped into the building. A heat pump is a device for transferring heat from a substance at one temperature to a substance at a different temperature by alternately vaporizing and liquefying a fluid with a compressor.

Better efficiency and wiser use of home appliances can play a major part in energy conservation. According to the Report of the Energy Project at the Harvard Business School, *Energy Future,* an accelerated effort to standardize and increase the efficiency of home appliances could lead to substantial savings without affecting people's standard of living.

Almost a third of residential energy is now consumed by home appliances. Consider the following list of approximate figures for various appliances that are used commonly. You may find that people who make statements about conserving energy by not using electric clocks are not contributing as much to conservation as they think they are. Using major appliances at times when there is a low load on utility lines can help.

COMMON APPLIANCES—ENERGY USE

The following are average monthly kilowatt hour (KWH) figures for common household appliances. By multiplying the KWH by the cost of a kilowatt of electricity in your area, you can estimate the monthly cost of each appliance. Your local electric company can supply this information.

Appliance	*Description of Usage*	*Estimated KWH/Month*
**Air conditioner*	*75 hr/month*	118
**Blanket, electric*	*8 hr/night*	18
Broiler	*8 hr/month*	11
Clock	*1 month*	1.5
Coffee percolator	*Twice daily*	9
Deep-fat fryer	*1hr/wk*	5
**Dehumidifier*	*1 month*	75
Dishwasher	*1 month*	30
Disposal	*1 hr/month*	.5
Dryer, clothes	*5 loads/wk*	80
Dryer, hair	*10 hr/month*	4
Fan, kitchen	*24 hr/month*	2
**Fan, window*	*1 month*	34
Food blender	*1 month*	1
Food freezer (15 cu. ft.)	*1 month*	100
Frying pan	*8 hr/month*	10
Grill, sandwich	*5 hr/month*	6
**Heating plant, burner*		
Motor	*1 month*	60
Hot air fan	*1 month*	80
Hot water circ.	*1 month*	30
**Heater, portable*	*40 hr/month*	53
Hotplate	*8 hr/month*	10
**Humidifier*	*1 month*	33
Iron, hand	*16 hr/month*	12
Lighting,		
5 rooms	*1 month*	70
7 rooms	*1 month*	80
10 rooms	*1 month*	100

Appliance	*Description of Usage*	*Estimated KWH/Month*
Microwave oven	*1 month*	25
Radio	*120 hr/month*	9
Range	*Family of 4*	100
Refrigerator,		
14 cu. ft., conventional	*1 month*	95
14 cu. ft., frost-free	*1 month*	150
Self-cleaning oven	*once a wk*	20
Sewing machine	*10 hr/month*	1
Shaver	*1 month*	2
Television,		
black & white	*180 hr/month*	30
color	*180 hr/month*	42
Vacuum cleaner	*4 hr/month*	3
Washing machine (automatic)	*5 loads/wk*	8
Water heater (quick recovery)	*1 month*	400
Water pump	*1 month*	20

* Peak use months

A number of industries have made large reductions in the amount of energy they use. Manufacturing eats up vast quantities of fuel and electricity, and conservation by industrial firms has been dramatic. It has ranged from simple actions, such as turning off motors that are not in use, to major changes, such as designing new methods of production that yield more product with less energy. Some firms report energy savings as high as 50 percent through conservation and a relatively small investment. But even though dramatic gains have been made, the potential for further conservation of energy by industry appears great.

"Cogeneration" is a form of conservation that seems to hold special promise. Cogeneration is the integration of the production of electricity and steam. In some systems, electricity is generated first, with steam as a byproduct. Waste steam is put to use in various heating and manufacturing processes. In other systems, steam is generated for and from industrial processes that can be used at lower temperatures and pressures to generate electricity. It has been estimated that industry in the United States could meet about half its own needs for electricity by 1985 through the increased use of cogeneration.

Cogeneration can help in residential heating systems to a limited degree. District heating is popular in Europe, where pipes from about a thousand power plants carry steam or hot water to homes and offices. This form of conservation is practical for some urban areas where large apartment and office buildings are heated this way.

The possible ways of conserving energy are numerous and varied. They range from changing the habits of inefficient cooks who use twice as much gas or electricity as necessary to large-scale industrial changes. Conservation is found on the lists of almost all the people and organizations that are making suggestions for the future.

Report after report suggests new mixes for the energy future. Some reports call for the increased use of coal and nuclear energy. Others put the emphasis on solar energy, calling for 25 percent of the supply from various solar sources by the year 2000 and 50 percent by the year 2020. Synthetic fuels are the hope of some energy planners. Few agree on the paths to follow, but none argues against conservation. Conservation is nonpolluting, inexpensive, involves no new technology, and is available now.

While experts argue the relative merits of energy sources, there is little quarrel about the seriousness of the fuel emergency. The Department of Energy in its Position Statement before the Committee on Science and Technology of the United States House of Representatives on January 31, 1980, reported that between now and 1985, the world will continue to rely heavily on oil, the source of half the world's energy in 1979. Decreases in United States oil and gas production will be exceeded by increased production from coal, nuclear, and renewable sources. Conservation is cited as the most readily available source of additional energy.

The report points out that, in the middle period

between 1985 and the year 2000, the world is expected to make a significant move away from dependence on oil. Options include more coal and coal-derived synthetic fuels, solar technology, oil shale, unconventional gas supplies, and nuclear power, as well as continued improvements in the efficiency of energy use.

Beyond the year 2000, the world will move further toward renewable resources and advanced nuclear technologies. These technologies will displace both traditional fuels and nonrenewable, unconventional sources of energy, but improvements in cost and technical performance must be achieved before they can be adopted widely.

The Department of Energy's report on the world energy outlook points out that, for many decades ahead, we must continue to make efforts to limit demand and to expand and diversify supply with equal diligence. The report states that No single energy source, no single restraint on demand, and no single technological innovation can resolve our current energy problems. Few disagree.

Certainly, more research and development of a wide variety of energy sources will help to increase the options for the long term. Experts who represent many viewpoints join the effort to promote awareness that the world's gauge will read closer to Empty without a large-scale and continued effort. One expert notes that it is difficult to be both realistic and hopeful, for he believes that no matter how much energy we conserve, we remain on the brink of catastrophe.

Whether or not conservation can play a major part in the transition from conventional fuels to renewable resources depends upon millions of people performing a wide variety of actions. You are one of those people.

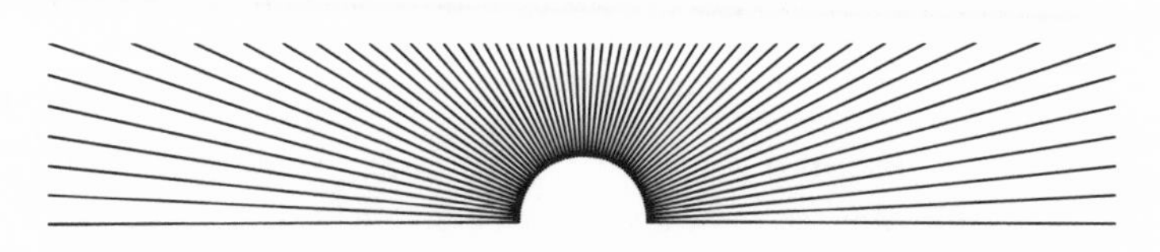

SUGGESTIONS FOR FURTHER READING

Blandy, Thomas and Denis Lamoureux, *All Through the House: A Guide to Home Weatherization.* New York: McGraw-Hill, 1980

Brown, Lester R., *Food or Fuel: New Competition for the World's Cropland.* Washington, D.C.: Worldwatch Institute, 1980

Brown, Lester R., Christopher Flavin, and Colin Norman, *The Future of the Automobile.* Washington, D.C.: Worldwatch Institute, 1979

Butti, Ken and John Perlin, *A Golden Thread: 2500 Years of Solar Architecture and Technology.* New York: Van Nostrand, Reinhold, 1980

Commoner, Barry, *The Politics of Energy.* New York: Knopf, 1979

Coonley, Douglas R. *Wind: Making It Work For You.* Philadelphia: The Franklin Institute Press, 1979

Crabbe, David and Richard McBride, eds., *The World Energy Book: An A–Z Atlas and Statistical Source Book.* Cambridge, Massachusetts: The MIT Press, 1979

Daniels, Farrington, *Direct Use of the Sun's Energy.* New York: Ballantine, 1974

Davis, Norah Deakin and Linda Lindsey, *At Home in the Sun: An Open House Tour of Solar Homes in the United States.* Charlotte, Vermont: Garden Way Publishing Company, 1979

DiCerto, Joseph J., *The Electric Wishing Well: The Solution to the Energy Crisis.* New York: Macmillan, 1976

Dorf, Richard C., *The Energy Factbook.* New York: McGraw-Hill, 1980

Goldstein, Jerome, *Recycling: How to Reuse Wastes in Home, Industry and Society.* New York: Schocken, 1979

Hayes, Denis, *Rays of Hope: The Transition to a Post-Petroleum World.* New York: Norton, 1979

Hoke, John, *Solar Energy* (rev. ed.). New York: Franklin Watts, 1978

Hyde, Margaret O. and Bruce G. Hyde, *Everyone's Trash Problem: Nuclear Wastes.* New York: McGraw-Hill, 1979

Kendall, Henry W. and Steven J. Nadis, eds., *Energy Strategies: Toward a Solar Future. A Report of the Union of Concerned Scientists.* New York: Ballinger, 1980

Landsberg, Hans H., Chairman, *Energy: The Next Twenty Years. A Report Sponsored by the Ford Foundation.* New York: Ballinger, 1979

Lovins, Amory B., *Soft Energy Paths: Toward a Durable Peace.* New York: Harper and Row, 1977

McPhillips, Martin and Bruce Anderson, eds., *The Solar Age Resource Book.* New York: Everest House, 1979

Mitz, William and Allen L. Hammond, *Solar Energy in America.* Washington, D.C.: American Association for the Advancement of Science, 1978

Moran, Joseph M., Michael D. Morgan, and James H. Wiersma, *Introduction to Environmental Science.* San Francisco: W. H. Freeman, 1980

Phillips, Owen, *The Last Chance Energy Book.* New York: McGraw-Hill, 1980

Readings from the Scientific American, Energy and Environment. San Francisco: W. H. Freeman, 1980

Schurr, Sam H., Joel Darmstadter, Harry Perry, William Ramsay and Milton Russell, *Energy in America's Future?* Published for Resources for the Future by Johns Hopkins University Press, Baltimore, 1979

Steinhart, John S., et al., *Pathway to Energy Sufficiency: The 2050 Study.* San Francisco: Friends of the Earth, 1979

Stobaugh, Robert and Daniel Yergin, *Energy Future: Report of the Energy Project at the Harvard Business School.* New York: Random House, 1979

Teller, Edward, *Energy from Heaven and Earth.* San Francisco: W. H. Freeman, 1979

Ward, Barbara, *Progress for a Small Planet.* New York: W. W. Norton, 1979

Wilson, Carroll L., *Coal: Bridge to the Future.* New York: Ballinger, 1980

FOR FURTHER INFORMATION

Many of the organizations or departments listed below provide free and/or low-cost materials that deal with energy. You may wish to write for their lists of publications and ask for any free materials dealing with a subject that interests you.

Alliance to Save Energy
1925 K Street N.W., Suite 507
Washington, D.C. 20006

* Especially important

American Association for the Advancement of Science
1776 Massachusetts Avenue N.W.
Washington, D.C. 20036

American Public Power Association
2600 Virginia Avenue N.W.
Washington, D.C. 20037

Committee for Nuclear Responsibility
Box 332
Yachats, Oregon 97498

Conservation Foundation
1717 Massachusetts Avenue N.W.
Washington, D.C. 20036

* Consumer Information Center
General Services Administration
Washington, D.C. 20405

* Energy
P. O. Box 62
Oak Ridge, Tennessee 37830

Energy Project
Center for Science in the Public Interest
1757 S Street N.W.
Washington, D.C. 20009

Environmental Action Reprint Service
2239 East Colfax
Denver, Colorado 80206

* National Solar Heating and Cooling Information Center
P.O. Box 1607
Rockville, Maryland 20805

* National Wildlife Federation
1412 Sixteenth Street N.W.
Washington, D.C. 20036

Office of Energy Conservation
Federal Energy Administration
Washington, D.C. 20461

Sierra Club
530 Bush Street
San Francisco, California 94108

United States Department of Agriculture
Publications Division
Office of Communication
Washington, D.C. 20205

* United States Department of Energy
Editorial Services Division
Office of Public Affairs
Mail Stop 8G-031
Washington, D.C. 20585

United States Environmental Protection Agency
Office of Public Affairs
401 M Street S.W.
Washington, D.C. 20460

* (*your*) State Department of Energy
State Office Building
(*your State Capital*)

DEPARTMENT OF ENERGY PUBLICATIONS

The following list is a selection of Department of Energy publications that help save energy. These or similar publications are available without charge from the Department of Energy Technical Information Center, P.O. Box 62, Oak Ridge, Tennessee 37830. If you are interested in another topic, ask if a free publication is available on that subject.

DOE/OPA-0013 **Energy from the Winds**
How the winds can be harnessed to generate electrical power.

DOE/OPA-0022 **Energy Savings Through Automatic Thermostat Controls**
Controls that save energy by setting back thermostats automatically.

EDM-1050 **Heat Pumps**
Heat pumps—operation, selection, and new developments.

DOE/OPA-0018 **How to Improve the Efficiency of Your Oil-Fired Furnace**
Beneficial effect on energy usage and the environment of regular servicing of oil-fired furnaces.

DOE/OPA-0021 **Insulate Your Water Heater and Save Fuel**
Value of insulating water heaters with a glass fiber insulation kit sold in local stores.

DOE/CS-0017 **Insulation**
Reducing energy waste in heating and cooling homes by installing adequate insulation.

DOE/CS-0006 **Lighting**
Energy-efficient uses of incandescent and fluorescent lamps.

DOE/OPA-0033 **Put the Sun to Work Today**
Solar energy—its applications, heating and cooling technologies, and factors to consider when buying home solar equipment.

DOE/OPA-0016 **Solar Energy**
Solar energy techniques for heating and cooling for agricultural and industrial processes and for using biomass and generating electricity.

DOE/OPA-0019 **Winter Survival**
A guide for saving energy and handling cold-weather emergencies.

ENERGY HOTLINES

Five energy hotline services are operated by the United States Department of Energy:

National Alcohol Fuels Information Center	800/525-5555
(Colorado only)	800/332-8339
Emergency Conservation Service	800/424-9122
(Alaska, Hawaii, Puerto Rico, Virgin Islands)	800/424-9088
(Washington, D.C., Area)	202/252-4950
Gasoline and Heating Oil Hotline	800/424-9246
(Washington, D.C., Area)	202/653-3437
Ridesharing Information	800/424-9184
(Washington, D.C., Area)	202/426-2943
Solar Heating and Cooling Information	800/523-2929
(Pennsylvania only)	800/462-4983

INDEX